2

Guidance Note 2

Isolation
& Switching

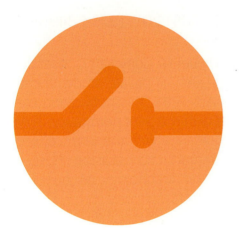

18th IET Wiring Regulations BS 7671:2018

Published by The Institution of Engineering and Technology, London, United Kingdom

The Institution of Engineering and Technology is registered as a Charity in England & Wales (no. 211014) and Scotland (no. SC038698).

The Institution of Engineering and Technology is the institution formed by the joining together of the IEE (The Institution of Electrical Engineers) and the IIE (The Institution of Incorporated Engineers).

© 1992, 1995, 1999, 2003 The Institution of Electrical Engineers
© 2009, 2012, 2015, 2019 The Institution of Engineering and Technology

First published 1992 (0 85296 536 2)
Reprinted (with minor amendments) 1993
Second edition (incorporating Amendment No. 1 to BS 7671:1992) 1995 (0 85296 866 3)
Third edition (incorporating Amendment No. 2 to BS 7671:1992) 1999 (0 85296 955 4)
Fourth edition (incorporating Amendment No. 1 to BS 7671:2001) 2003 (0 85296 990 2)
Reprinted (incorporating Amendment No. 2 to BS 7671:2001) 2004
Fifth edition (incorporating BS 7671:2008) 2009 (978-0-86341-856-3)
Sixth edition (incorporating Amendment No. 1 to BS 7671:2008) 2012 (978-1-84919-273-6)
Reprinted (with minor editorial amendments) 2012
Reprinted 2014
Seventh edition (incorporating Amendment Nos. 2 and 3 to BS 7671:2008) 2015 (978-1-84919-871-4)
Eighth edition (incorporating BS 7671:2018) 2018 (978-1-78561-449-1)

Copies of this publication may be obtained from:
The Institution of Engineering and Techology
PO Box 96, Stevenage, SG1 2SD, UK
Tel: +44 (0)1438 767328
Email: sales@theiet.org
www.theiet.org/wiringbooks

ISBN 978-1-78561-449-1 (paperback)

ISBN 978-1-78561-450-7 (electronic)

Typeset in the UK by the Institution of Engineering and Technology
Printed in the UK by Sterling Press Ltd, Kettering

Contents

Cooperating organisations

The Institution of Engineering and Technology acknowledges the invaluable contribution made by the following individuals in the preparation of this Guidance Note.

M. Coles BEng (Hons) MIET
Eur Ing L. Markwell MSc BSc (Hons) CEng MIET MCIBSE LCGI

We would like to thank the following organisations for their continued support:

BEAMA Installation
Certsure trading as NICEIC and Elecsa
ECA
SELECT
Health and Safety Executive
NAPIT
Stroma Certification

Acknowledgements

References to British Standards, CENELEC Harmonization Documents and International Electrotechnical Commission (IEC) standards are made with the kind permission of BSI. Complete copies can be obtained by post from:

BSI Customer Services 389 Chiswick High Road London W4 4AL
Tel: +44 (0)20 8996 9000 Fax: +44 (0)20 8996 7001
Email: cservices@bsigroup.com

BSI also maintains stocks of international and foreign standards, with many English translations.Up-to-date information on BSI standards can be obtained from the BSI website: www.bsigroup.com

Advice is included with the kind permission of the Energy Networks Association Limited. Complete copies of their publications can be obtained by post from:

Energy Networks Association
6th Floor, Dean Bradley House
52 Horseferry Road
London SW1P 2AF
Tel: +44 (0)20 7706 5100.
Email: info@energynetworks.org

Documents available from their website www.energynetworks.org include Technical Specifications, BEBSSpecifications, Engineering Recommendations and a variety of reports.

Copies of Health and Safety Executive documents and approved codes of practice (ACOP) are available free of charge on the HSE website (www.hse.gov.uk) and can be obtained from:

HSE Books PO Box 1999 Sudbury, Suffolk CO10 2WA Tel: +44 (0)1787 881165Email: hsebooks@prolog.uk.com Web: http://books.hse.gov.uk

The illustrations within this publication were provided by Farquhar Design: www.farquhardesign.co.uk

Cover design and illustration were created by The Page Design: www.thepagedesign.co.uk

Preface

This Guidance Note is part of a series issued by the Institution of Engineering and Technology to explain and enlarge upon the requirements in BS 7671:2018, the 18th Edition of the IET Wiring Regulations. All references to BS 7671 in the text of this Guidance Note are references to the current 18th Edition of BS 7671 unless otherwise noted.

This Guidance Note is intended to explain and illustrate some of the requirements of BS 7671, but users should always consult BS 7671 to satisfy themselves of compliance with its requirements.

The scope generally follows that of BS 7671:2018; the relevant Regulations and Appendices are noted in the margin. Some Guidance Notes also contain material not included in BS 7671:2018 but which was included in earlier editions of the Wiring Regulations. All of the Guidance Notes contain references to other relevant sources of information.

Electrical installations in the United Kingdom that are designed and constructed to comply with BS 7671 are likely to satisfy the relevant parts of Statutory Regulations such as The Electricity at Work Regulations 1989. However, this cannot be guaranteed. It is stressed that it is essential to establish which Statutory and other Regulations apply and to design and construct installations accordingly. For example, an installation in a petrol filling station or in premises subject to licensing may have requirements different from or additional to, those of BS 7671 and these will take precedence.

For a supply in the UK given in accordance with The Electricity Safety, Quality and Continuity Regulations or ESQCR (N Ireland), it can be assumed that the connection with Earth of the neutral of the supply is permanent. Outside England, Scotland, Northern Ireland and Wales, confirmation must be sought from the distributor that the supply conforms to requirements corresponding to those of The Electricity Safety, Quality and Continuity Regulations 2002, in this respect. Where the ESQCR do not apply, equipment for isolation and switching must be selected and installed accordingly.

Electrical installations must be maintained during their working life for safety, and periodic inspection and testing is advised during the life of the installation to assist in identifying damage, deterioration, defects or dangerous conditions and maintenance requirements. See IET Guidance Note 3 for further details. It is the duty of the building's or site "duty holder" to maintain their installation so it is safe and to undertake such maintenance so this can be achieved. Periodic inspection and testing can assist in this maintenance.

Introduction

This Guidance Note is concerned primarily with Chapter 46 and Section 537 and those other parts of BS 7671 which relate to isolation, switching and control including provision of information and labelling. Requirements relating to isolation and switching in particular types of installation and for specific items of installed equipment are also covered.

Neither BS 7671 nor the Guidance Notes are design guides. It is essential to prepare a full design and specification prior to commencement or alteration of an electrical installation. Compliance with the relevant legislation (such as The Electricity at Work Regulations) is necessary, and compliance with relevant standards should be required (see Regulations 132.1, 133.1 and 511.2). Precise details of each item of equipment should be obtained from the manufacturer and/or supplier and compliance with appropriate standards confirmed.

The design and specification documents should set out the requirements and provide sufficient information to enable an electrically skilled person or persons to carry out the installation and to commission it. The specification must include a description of how the system is to operate and all the design and operational parameters. It must provide for all the commissioning procedures that will be required and for the provision of adequate operation and maintenance information to the user. This will be by means of an operational and maintenance manual or schedule, and 'as fitted' drawings if necessary.

It must be noted that it is a matter of contract as to which person or organisation is responsible for the production of the parts of the design, specification, construction and verification of the installation and any operational information.

The persons or organisations who may be concerned in the preparation of the works include:

The Designer
The CDM Principal Designer and Principal Contractor
The Contractor or Installer
The Verifier
Specialist Commissioning Engineers
The Electricity Supplier or Distributor Network Operator (DNO or other supplier)
The Installation Owner (Client) and/or User
The Architect
The Fire Risk Assessor (a competent person for building inspection for fire risk and safety)
Specialist manufacturers or suppliers
Any Regulatory Authorities
Local Building Control Authority / Standards Division or Approved Inspector
Any Specialist Insurers
The Health and Safety Executive

© The Institution of Engineering and Technology

In producing the design, advice should be sought from the installation owner and/or user as to the intended use and operational requirements. Often, as in a speculative building, the intended user is unknown but the proposed type of building – office, workshop, accommodation etc. – must be established. The specification must provide for all the commissioning procedures that will be required and for the production of any operational and maintenance manual interfacing with the building logbook.

The operational and maintenance manual must include a description of how the system as installed is to operate, maintenance provisions and all test and commissioning records. The manual should also include manufacturers' technical data for all items of switchgear (including any protection settings or device data), luminaires, accessories, etc. maintenance guidance and any special instructions that may be needed.

The Health and Safety at Work etc. Act 1974 Section 6 and The Construction (Design and Management) Regulations 2015 are concerned with the provision of information, and guidance on the preparation of *technical manuals* is given in BS EN 82079-1 *Preparation of instructions for use. Structuring, content and presentation. General principles and detailed requirements* and the BS 4940 series *Technical information on constructional products and services*. The size and complexity of the installation will dictate the nature and extent of the manual.

The IET/IEE Wiring Regulations have been extensively referred to in HSE guidance over the years. Installations which conform to the standards laid down in BS 7671:2018 are regarded by HSE as likely to achieve conformity with the relevant parts of The Electricity at Work Regulations 1989. Existing installations may have been designed and installed to conform to the standards set by earlier editions of BS 7671 or the IEE Wiring Regulations. This does not mean that they will fail to achieve conformity with the relevant parts of The Electricity at Work Regulations 1989 but it is for a competent person or persons to assess the installation and take an engineering view as to its compliance or otherwise (see IET Guidance Note 3 for further guidance).

Before any work begins a work programme must be agreed with the site 'duty-holder'/ client and a Health and Safety plan including risk assessments and method statements agreed and put in place. It may be that work has to be carried out when the premises is operating and staff are present so they must not be exposed to any safety risks. Provide effective control of any area where work is being undertaken, for example, the work areas should be fenced off with temporary barriers and electrical equipment not left unattended when open. Escape routes must be kept open at all times or temporary alternatives arranged. Where supplies have to be isolated it is necessary that this must be planned in advance with the premises management. It is the duty of the buildings or site 'duty-holder' to maintain their installation so it is safe and to undertake such maintenance so this can be achieved.

Statutory requirements

1.1 General

A number of enactments and statutory instruments including The Electricity at Work Regulations 1989 (made under the Health and Safety at Work etc. Act 1974) are of relevance to isolation and switching. These include machinery safety requirements which also come under the Supply of Machinery (Safety) Regulations 2008, with the 2011 amendment.

110.2 This Guidance Note is not intended to provide an exhaustive treatment of the legislation concerned with isolation and switching in low voltage installations, but deals only with those situations referred to in BS 7671. Thus, certain specialised installations listed in Regulation 110.2 are excluded.

1.2 Statutory regulations

1.2.1 The Health and Safety at Work etc. Act 1974

The Health and Safety at Work etc. Act 1974, Part I, Section 6, places a duty on any person ("person" includes the plural or any company) who designs, manufactures, imports or supplies any article for use at work to ensure that adequate information is provided so that when put to that use, it will be safe and without foreseeable risks to health. Schedule 3 of the Consumer Protection Act 1987 extended the use to when the equipment is being set, used, cleaned or maintained.

Those with duties under the Act should include in their written instructions, manuals, etc. details of the means of isolation and other health and safety provisions provided and the need to use these when and where required.

An electrical installation may well, under some circumstances, be considered to be 'an article for use at work' and there is a duty on the designer and installer to provide adequate instruction and/or maintenance manuals irrespective of any contract provisions.

1.2.2 The Electricity at Work Regulations 1989

The Electricity at Work Regulations (EWR) 1989 are general in their application and refer throughout to 'danger' and 'injury'. Danger is defined as risk of 'injury' and injury is defined in terms of certain classes of potential harm to persons. Injury is stated to mean death or injury to persons from:

▶ electric shock
▶ electric burn
▶ electrical explosion or arcing
▶ fire or explosion initiated by electrical energy.

Of particular interest to the subject matter of this publication are Regulation 12 'Means for cutting off the supply and for isolation' and Regulation 13 'Precautions for work on equipment made dead'.

Regulation 12(1)(b) states that where necessary to prevent danger, suitable means shall be available for the isolation of any electrical equipment, where isolation means the disconnection and separation of the electrical equipment from every source in such a way that the disconnection and separation is secure (para (2) refers).

The *Electricity at Work Regulations 1989. Guidance on Regulations (HSR)25* published by the Health and Safety Executive advises with reference to Regulation 12(1)(b) above that isolation is the process of ensuring that the supply to all or a particular part of an installation remains isolated and that inadvertent reconnection is prevented.

The coverage of Regulation 12 and 13 in HSR25 highlights the need to enable the prevention of unauthorised, improper or unintentional energisation and provides an example of locking-off facilities. This aligns with BS 7671 Regulation 462.3 which requires devices for isolation to be designed and/or installed so as to prevent unintentional or inadvertent closure. Examples of precautions are as follows:

▶   Located within a lockable space or lockable enclosure
▶   Padlocking
▶   Located adjacent to the associated equipment

Precautions for work on equipment made dead

Adequate precautions shall be taken to prevent electrical equipment, which has been made dead in order to prevent danger while work is being carried out on or near that equipment, from becoming electrically charged during that work if danger may thereby arise.

The coverage of Regulation 13 in HSR25 highlights the need to lock off or otherwise secure any device being used to provide isolation. Where isolation has been achieved by the removal of fuses these can provide a secure arrangement by their retention in a safe place if proper control procedures are used to prevent unauthorized or inadvertent access to the fusegear during the isolation period.

The *The Electricity at Work Regulations 1989. Guidance on Regulations (HSR)25* is essential reading for all concerned with electrical installations.

The Electricity at Work Regulations generally considers that "live working" (working with the supply on) is dangerous and not acceptable. There may be occasions where it could be acceptable; but these are quite unlikely and Regulation 14 of The Electricity at Work Regulations must be considered.

BS 7671 is intended primarily for designers and installers and thus covers only the provision of isolators and the means of securing them. The responsibility for ensuring that equipment is properly isolated when necessary remains with the user.

In HSR 25 (see the reference to Regulation 12(1) (a)) it is recognized that there may be a need to switch off electrical equipment for reasons other than preventing electrical danger, but these considerations are outside the scope of The Electricity at Work Regulations 1989.

131.1(v) BS 7671, however, takes into account injury from mechanical movement of electrically actuated equipment. Emergency switching and switching off for mechanical maintenance are therefore included in BS 7671.

The Electricity at Work Regulations 1989 specifically require adequate suitability and functionality of switches and isolators as well as maintenance, and this implies inspection of electrical systems, supplemented by testing as necessary. Regular functional testing of safety circuits such as emergency switching/stopping, etc. may be required – especially if they are very rarely required to be used. Comprehensive records of all inspections and tests should be made and reviewed for any trends that may arise. Guidance Note 3: *Inspection & Testing* gives more detailed guidance on initial and periodic inspection and testing of installations.

In Regulation 12(3) of The Electricity at Work Regulations it is noted that there are certain pieces of electrical equipment which are themselves a source of electrical energy, such as batteries or photo-voltaic equipment, and in such a case as is necessary precautions must be taken to prevent danger so far as is reasonably practicable.

1.2.3 The Electricity Safety, Quality and Continuity Regulations

The prime purpose of the Electricity Safety, Quality and Continuity Regulations (ESQCR) is to provide for the safety of the public and to ensure an electricity supply of adequate quality and reliability. Initially these regulations were implemented in 2002 and applied only to England, Scotland and Wales. Northern Ireland did not implement the ESQCR text until 2008. The ESQCR is usually quoted without a date to show that it covers all of the UK.

These regulations make reference to BS 7671. For example, under Regulation 25(2) the consumer may have to satisfy the local electricity distributor that the electrical installation is safe and technically sound by providing evidence that it complies with BS 7671. An Electrical Installation Certificate would normally be acceptable.

Under Regulation 28 of the ESQCR a distributor must provide, in respect of any existing or proposed consumer's installation which is connected or is to be connected to the distributors network, to any person who can show a reasonable cause for requiring the information, a written statement of:

(a) the maximum prospective short circuit current at the supply terminals;
(b) the maximum earth loop impedance of the earth fault path outside the installation (for low voltage connections);
(c) the type and rating of the distributor's protective device or devices nearest to the supply terminals;
(d) the type of earthing system applicable to the connection; and
(e) before commencing a supply to a consumer's installation, or when the existing supply characteristics have been modified, the supplier shall ascertain from the distributor and then declare to the consumer –
  (i) the number of phases;
  (ii) the frequency; and
  (iii) the voltage
at which it is proposed to supply electricity, which apply, or will apply, to that installation.

1.2.4 The Machinery Directive

The EU Machinery Directive (EU Directive, 2066/42/EC) was implemented into UK law by The Supply of Machinery (Safety) Regulations 2008 and The Supply of Machinery (Safety) (Amendment) Regulations 2011. The Directive aims to provide greater clarity than the old directive, for example, with a modified definition of the core concept of 'machinery' and in the dividing lines between itself and the Lifts and Low Voltage Directives. These Regulations introduce a quality assurance module as a conformity assessment option for relevant manufacturers.

In the Supply of Machinery (Safety) Regulations, machinery is defined as:

(i) an assembly, fitted with or intended to be fitted with a drive system other than directly applied human or animal effort, consisting of linked parts or components, at least one of which moves, and which are joined together for a specific application;

(ii) an assembly as referred to in sub-paragraph (i), missing only the components to connect it on site or to sources of energy and motion;

(iii) an assembly as referred to in sub-paragraph (i) or (ii), ready to be installed and able to function as it stands only if mounted on a means of transport, or installed in a building or structure;

(iv) assemblies of machinery as referred to in sub-paragraphs (i), (ii) and (iii) or partly completed machinery, which, in order to achieve the same end, are arranged and controlled so that they function as an integral whole;

(v) an assembly of linked parts or components, at least one of which moves and which are joined together, intended for lifting loads and whose only power source is directly applied human effort; and

(vi) they were placed on the market or put into service on or after 29th December 2009.

The Regulations do not apply:

(a) to machinery specific to specialist equipment listed in Schedule 3 of the Regulations, including ships and offshore platforms for which reference should be made to the relevant statutory instrument

(b) to machinery previously used in the European Community

(c) to machinery for use outside the European Economic Area

(d) where the risks are mainly of electrical origin (such machinery is covered by The Electrical Equipment (Safety) Regulations 2016) – see Section 1.2.5.

Note: These Statutory Regulations apply only to low voltage equipment up to 1 kV

(e) where the risks are wholly or partly covered by other Directives, from the date those other Directives are implemented into United Kingdom law

(f) to machinery first supplied in the European Community before 1 January 1993.

Machinery manufactured in conformity with specified published European standards that have also been listed in the *Official Journal of the European Communities* will be presumed to comply with the essential health and safety requirements of those standards and hence the regulations.

BS EN 60204 *Safety of Machinery. Electrical equipment of machines* is the major standard on machine electrical equipment. BS EN ISO 12100 *Safety of machinery. General principles for design. Risk assessment and risk reduction* and BS EN ISO 13850 *Safety of machinery. Emergency stop equipment, functional aspects. Principles for design* also give advice on emergency switching.

Annex I to the Supply of Machinery (Safety) Regulations 2008 provides essential advice. The advice on general machinery control is reproduced here as follows:

1.2.4.1 CONTROL SYSTEMS

1.2.4.1.1 Safety and reliability of control systems

Control systems must be designed and constructed in such a way as to prevent hazardous situations from arising. Above all, they must be designed and constructed in such a way that:

- they can withstand the intended operating stresses and external influences,
- a fault in the hardware or the software of the control system does not lead to hazardous situations,
- errors in the control systems' logic do not lead to hazardous situations,
- reasonably foreseeable human error during operation does not lead to hazardous situations.

Particular attention must be given to the following points:

- the machinery must not start unexpectedly,
- the parameters of the machinery must not change in an uncontrolled way, where such change may lead to hazardous situations,
- the machinery must not be prevented from stopping if the stop command has already been given,
- no moving part of the machinery or piece held by the machinery must fall or be ejected,
- automatic or manual stopping of the moving parts, whatever they may be, must be unimpeded,
- the protective devices must remain fully effective or give a stop command,
- the safety-related parts of the control system must apply in a coherent way to the whole of an assembly of machinery and/or partly completed machinery.

For cable-less control, an automatic stop must be activated when correct control signals are not received, including loss of communication.

1.2.4.1.2 Control devices

Control devices must be:

- clearly visible and identifiable, using pictograms where appropriate,
- positioned in such a way as to be safely operated without hesitation or loss of time and without ambiguity,
- designed in such a way that the movement of the control device is consistent with its effect,
- located outside the danger zones, except where necessary for certain control devices such as an emergency stop or a teach pendant,
- positioned in such a way that their operation cannot cause additional risk,
- designed or protected in such a way that the desired effect, where a hazard is involved, can only be achieved by a deliberate action,
- made in such a way as to withstand foreseeable forces; particular attention must be paid to emergency stop devices liable to be subjected to considerable forces.

Where a control device is designed and constructed to perform several different actions, namely where there is no one-to-one correspondence, the action to be performed must be clearly displayed and subject to confirmation, where necessary.

Control devices must be so arranged that their layout, travel and resistance to operation are compatible with the action to be performed, taking account of ergonomic principles.

Machinery must be fitted with indicators as required for safe operation. The operator must be able to read them from the control position.

From each control position, the operator must be able to ensure that no-one is in the danger zones, or the control system must be designed and constructed in such a way that starting is prevented while someone is in the danger zone.

If neither of these possibilities is applicable, before the machinery starts, an acoustic and/or visual warning signal must be given. The exposed persons must have time to leave the danger zone or prevent the machinery starting up.

If necessary, means must be provided to ensure that the machinery can be controlled only from control positions located in one or more predetermined zones or locations.

Where there is more than one control position, the control system must be designed in such a way that the use of one of them precludes the use of the others, except for stop controls and emergency stops.

When machinery has two or more operating positions, each position must be provided with all the required control devices without the operators hindering or putting each other into a hazardous situation.

1.2.5 The Electrical Equipment (Safety) Regulations 2016

Subject to paragraphs (a) and (b) below, these regulations apply to any electrical equipment (including any electrical apparatus or device) designed or adapted for use with voltage (in the case of alternating current) of not less than 50 V nor more than 1,000 V or (in the case of direct current) of not less than 75 V nor more than 1,500 V.

(a) The regulations do not apply to the following types of electrical equipment:
  (i) Equipment for use in an explosive atmosphere
  (ii) Equipment for radiology and medical purposes
  (iii) Parts for goods lifts and passenger lifts
  (iv) Electricity meters
  (v) Plugs and socket outlets for domestic use
  (vi) Electric fence controllers
  (vii) Custom-built evaluation kits destined for professionals to be used at research and development facilities solely for research and development.
  (viii) Specialised electrical equipment for use on ships, aircraft or railways, which complies with the safety provisions drawn up by international bodies in which the member States participate.
(b) The regulations do not apply to any electrical equipment supplied for export to a place which is not within any member state.

The regulations do not apply to any electrical equipment which is placed on the market before 1st January 1997 and which complies with the provisions of the Low Voltage Electrical Equipment (Safety) Regulations 1989.

As noted in Section 1.2.4, certain machines are excluded from the Supply of Machinery (Safety) Regulations, where these have risks that are mainly of electrical origin and such equipment is covered by the Electrical Equipment (Safety) Regulations 2016.

Note: The safety requirements for placing products on the market are complex and Sections 1.2.4 and 1.2.5 above are only intended as a brief outline. Persons intending to develop products are advised to obtain specialist advice and guidance at an early stage in the design process.

1.2.6 The Construction (Design and Management) Regulations 2015

The Construction (Design and Management) Regulations (CDM Regulations) require active planning, coordination and management of the building works, including the electrical installation, to ensure that hazards associated with the construction, maintenance and perhaps demolition of the installation are given due consideration and designed out where possible, as well as provision for safety in normal maintenance use. The CDM Regulations do not apply to the day to day normal use of a building.

The regulations were extensively revised and reissued in 2015 and specific duties were identified for specific roles, and are shown on the following pages:

CDM duty-holders: who are they?	Summary of role/main duties
Clients are organisations or individuals for whom a construction project is carried out.	Make suitable arrangements for managing a project. This includes making sure: **(a)** other duty-holders are appointed; and **(b)** sufficient time and resources are allocated. Make sure: **(a)** relevant information is prepared and provided to other duty-holders; **(b)** the principal designer and principal contractor carry out their duties; and **(c)** welfare facilities are provided.
Domestic clients are people who have construction work carried out on their own home, or the home of a family member that is **not** done as part of a business, whether for profit or not.	Domestic clients are in scope of CDM 2015, but their duties as a client are normally transferred to: **(a)** the contractor, on a single contractor project, or; **(b)** the principal contractor, on a project involving more than one contractor. However, the domestic client can choose to have a written agreement with the principal designer to carry out the client duties.
Designers are those, who as part of a business, prepare or modify designs for a building, product or system relating to construction work.	When preparing or modifying designs, to eliminate, reduce or control foreseeable risks that may arise during: **(a)** construction; and **(b)** the maintenance and use of a building once it is built. Provide information to other members of the project team to help them fulfil their duties.
Principal designers are designers appointed by the client in projects involving more than one contractor. They can be an organisation or an individual with sufficient knowledge, experience and ability to carry out the role. (Principal designers are **not** a direct replacement for CDM co-ordinators. The range of duties they carry out is different to those undertaken by CDM co-ordinators under CDM 2007).	Plan, manage, monitor and coordinate health and safety in the pre-construction phase of a project. This includes: **(a)** identifying, eliminating or controlling foreseeable risks; and **(b)** ensuring designers carry out their duties. Prepare and provide relevant information to other duty-holders. Provide relevant information to the principal contractor to help them plan, manage, monitor and coordinate health and safety in the construction phase.

CDM duty-holders: who are they?	Summary of role/main duties
Principal contractors are contractors appointed by the client to coordinate the construction phase of a project where it involves more than one contractor.	Plan, manage, monitor and coordinate health and safety in the construction phase of a project. This includes: **(a)** liaising with the client and principal designer; **(b)** preparing the construction phase plan; and **(c)** organising cooperation between contractors and coordinating their work. ensure: **(a)** suitable site inductions are provided; **(b)** reasonable steps are taken to prevent unauthorised access; **(c)** workers are consulted and engaged in securing their health and safety; and **(d)** welfare facilities are provided.
Contractors are those who do the actual construction work and can be either an individual or a company.	Plan, manage and monitor construction work under their control so that it is carried out without risks to health and safety. For projects involving more than one contractor, coordinate their activities with others in the project team – in particular, comply with directions given to them by the principal designer or principal contractor. For single-contractor projects, prepare a construction phase plan.
Workers are the people who work for or under the control of contractors on a construction site.	They must: **(a)** be consulted about matters which affect their health, safety and welfare; **(b)** take care of their own health and safety and others who may be affected by their actions; **(c)** report anything they see which is likely to endanger either their own or others' health and safety; and **(d)** cooperate with their employer, fellow workers, contractors and other duty-holders.

Table reproduced from HSE publication L153

Regulation 12 requires that during the pre-construction phase a construction phase health and safety file is to be prepared, and this file must be reviewed and updated during the construction process. The construction phase plan must set out the arrangements for securing health and safety during the period construction work is carried out.

Regulation 12 also requires that during the pre-construction phase the principal designer must prepare a health and safety file appropriate to the characteristics of the project which must contain information relating to the project which is likely to be needed during any subsequent project to ensure the health and safety of any person. During the project, the principal contractor must provide the principal designer with any information in the principal contractor's possession relevant to the health and safety file, for inclusion in the health and safety file.

Appendices 2 and 3 of HSE publication L153 *"Managing health and safety in construction"* Construction (Design and Management) Regulations 2015 provides more details on the requirements for the provision of pre-construction phase and construction phase information and health and safety plans.

1.2.7 The Management of Health and Safety at Work Regulations 1999

These regulations were introduced to reinforce the Health and Safety at Work Act 1974 and place general duties on employers to assess risks to the health and safety of employees and others who may be affected by their work activity, and take managerial action to minimise these risks, including:

▶ implementing preventive measures
▶ providing health surveillance
▶ appointing competent people
▶ setting up procedures
▶ providing information
▶ training, etc.

By its very nature the risk assessment must include operation of equipment and machines, and the safety of the fixed installation, and must cover the adequate provision of isolation and emergency switching and stopping devices suitable for the considered risk. It should also cover access interlocks and any other equipment safety operating provisions such as guards, etc.

The risk assessment process may result in the need for a safe system of work or method statement to be produced. The statement need be no longer than necessary to achieve the objective of safe working and may include reference to relevant safety rules and permit to work procedures.

The general requirement is to identify and design to reduce risks in order to protect against hazards and dangers.

1.2.8 Workplace (Health, Safety and Welfare) Regulations 1992

These regulations cover a wide range of basic health, safety and welfare issues and apply to most workplaces except those involving construction work on construction sites, those in or on a ship, or those below ground at a mine.

The workplace and the equipment, devices and systems to which these regulations apply must be maintained in an efficient state, in efficient working order and in good repair, including cleaning as appropriate. Adequate heating, ventilation and lighting are required in all working areas including switchrooms, workshops etc., and around plant and machinery. Equipment that could fail and put workers in danger should be properly maintained and checked at regular intervals, as appropriate, by inspection, testing, adjustment, lubrication, repair and cleaning. The frequency of regular maintenance, and precisely what it involves and who is competent to complete it, will depend on the equipment or device concerned. There is guidance available from the HSE and advice from other authoritative sources, particularly manufacturers' information and instructions, as well as competent persons.

1.2.9 The Provision and Use of Work Equipment Regulations 1998

These regulations require employers to ensure that work equipment is suitable for the purpose and that users are properly trained in its use. Work equipment is defined in the regulations as being any machine, equipment or tool or installation used at work, whether exclusively or not.

The regulations also require that work equipment is inspected as necessary and maintained in an efficient state, in efficient working order and in good repair.

Regulations 14 to 18 cover controls and control systems, Regulation 19 details isolation requirements and Regulation 21 requires the provision of suitable and sufficient lighting for safe operation and working. Regulations 5, 6 and 22 are concerned with equipment maintenance.

1.2.10 The Regulatory Reform (Fire Safety) order 2005

This was probably the single most influential statutory instrument in the field of protection against fire in recent years. It has had a direct influence on other pieces of primary and secondary statutory legislation, requiring modifications to and in some cases partial or full revocation of the requirements therein. It replaced fire certification under the Fire Precautions Act 1971 with a general duty to ensure, so far as is reasonably practicable, the safety of employees and a general duty, in relation to non-employees, to take such fire precautions as may reasonably be required to ensure that premises are safe. It also requires a risk assessment to be carried out and regularly updated.

The Regulatory Reform (Fire Safety) Order 2005 (henceforth referred to as the Order) came into effect fully in October 2006 and affects over 70 pieces of fire safety law, not all of which fall within the scope of this Guidance Note.

The Order applies to non-domestic premises in England and Wales, including any communal areas of blocks of flats or houses in multiple occupation, and places legal obligations and responsibilities upon the *responsible person* defined in article 3 of the Order as follows:

---

3. Meaning of 'responsible person'

In this Order 'responsible person' means –

(a) in relation to a workplace, the employer, if the workplace is to any extent under his control;
(b) in relation to any premises not falling within paragraph (a) –
    (i) the person who has control of the premises (as occupier or otherwise) in connection with the carrying on by him of a trade, business or other undertaking (for profit or not); or
    (ii) the owner, where the person in control of the premises does not have control in connection with the carrying on by that person of a trade, business or other undertaking.

(This person can also be considered as the "duty-holder" under health and safety legislation)

Part 2 of the Order 'Fire Safety Duties' requires the responsible person to provide the necessary general fire precautions and training to ensure, so far as is reasonably practicable, the safety of employees and other persons within the premises for which they are responsible. The responsible person must carry out (or have carried out by a competent person) a fire safety risk assessment on the premises to identify what general fire precautions are required for the above to be achieved. The risk assessment must be regularly reviewed by the responsible person to keep it up to date and any fire safety requirements implemented.

The significant findings of the risk assessment including any actions that have been taken or which will be taken by the responsible person and details of any persons identified as being especially at risk must be recorded as soon as practicable afterwards where:

▶ the responsible person employs five or more persons, or
▶ the premises to which the assessment relates are subject to a licensing arrangement, or
▶ the premises are subject to an alterations notice.

Article 23 requires that every employee must, whilst at work:

(a) take reasonable care for the safety of themselves and of other persons who may be affected by their acts or omissions at work;

(b) as regards any duty or requirement imposed on their employer by or under any provision of the Order, cooperate with them so far as is necessary to enable that duty or requirement to be performed or complied with; and

(c) inform their employer or any other employee with specific responsibility for the safety of fellow employees –

   (i) of any work situation which they would reasonably consider represented a serious and immediate danger to safety; and

   (ii) of any matter which they would reasonably consider represented a shortcoming in the employer's protection arrangements for safety,

in so far as that situation or matter either affects the safety of the employee or arises out of or in connection with their activities at work, and has not previously been reported to the employer.

Article 37 relates to firefighters' switches for luminous tube signs, etc. This is a topic dealt with in greater depth in Chapter 8 of this Guidance Note.

It should be noted that in Scotland the Fire (Scotland) Act 2005 provides similar requirements for fire safety duties to those of the Regulatory Reform (Fire Safety) Order 2005 applicable to England and Wales. In Northern Ireland, The Fire Safety Regulations (Northern Ireland) 2010 apply.

1.2.11 Building Regulations Approved Document M

For new buildings, Approved Document M requires sockets, switches and controls to be mounted 450 mm–1.2 m above finished floor level. Installers may wish to adopt these mounting heights in agreement with their client for existing buildings.

Approved Document M also requires consumer units to be mounted so that the switches are 1350-1450 mm above floor level. This is to enable people with reduced reach to be able to access the main switch, circuit breakers and to carry out routine testing of any RCDs or AFDDs.

Overview of the Wiring Regulations

2

BS 7671:2018 *Requirements for Electrical Installations* IET Wiring Regulations Eighteenth Edition was published in July 2018 and runs in parallel with BS 7671:2008 Amendment 3:2015 until the 31st December 2018. During this parallel time both sets of regulations are valid but designers starting new projects are advised to start them using BS 7671:2018.

Through this Guidance Note all references to "BS 7671" or "The Regulations" are to be taken to imply BS 7671:2018 unless otherwise stated.

Within the Regulations:

Part 1 sets out the scope, object and fundamental principles.

Part 2 defines the sense in which certain terms are used throughout the Regulations, and provides a list of symbols used and a list of abbreviations used in the Standard. The subjects of the subsequent parts are as indicated below:

Part 3 Identification of the characteristics of the installation that will need to be taken into account in choosing and applying the requirements of the subsequent parts. These characteristics may vary from one part of an installation to another and should be assessed for each location to be served by the installation.

Part 4 Description of the measures that are available for the protection of persons, livestock and property, and against the hazards that may arise from the use of electricity.

Part 5 Precautions to be taken in the selection and erection of the equipment of the installation.

Part 6 Inspection and testing.

Part 7 Special installations or locations – particular requirements.

With respect to "Isolation and Switching":

▶ Chapter 13 prescribes the fundamental principles for safety and includes a number of regulations that are directly concerned with isolation and switching and sets the scene for all remaining parts of the Standard.
▶ Part 2 contains definitions specific to isolation and switching.
▶ Chapter 46 Isolation and Switching provides details of the available isolation and switching measures.

Sect 537 Section 537 Isolation and Switching provides details of the application of these measures.

131.7
131.8 Regulation 131.7 requires consideration to be given to the effects on the installation and equipment installed therein caused by an interruption of the supply.

132.9
132.10
132.12
132.13
132.15 Regulation 132.8 requires consideration of a circuits operating values and the suitability of the characteristics of protective equipment (switchgear etc.). Regulation 132.9 states the fundamental requirement that any emergency control device be so installed so that it is easily recognized and effectively and rapidly operated in the case of danger arising. Regulations 132.10 and 132.15 requires the suitable provision of disconnecting devices to permit the safe use of the installation in terms of operation, inspection, fault detection, diagnosis and repair, routine testing, and maintenance. There is also a specific requirement to provide a means of switching off every electric motor in an installation. Although not specific to isolation and switching alone, the accessibility requirements of Regulation 132.12 should be borne in mind when considering suitable location for items of switchgear within an installation. This is also true of the requirements of Regulation 132.13 regarding the provision of suitable documentation for the installation.

132.14.1
132.14.2 Regulation 132.14.1 makes it clear that single-pole devices including fuses, switches and circuit-breakers may only be installed in the line conductor of a circuit, while Regulation 132.14.2 prohibits the installation of such single-pole devices in an earthed neutral conductor.

133.1 Regulation group 133.1 requires every item of equipment used in an electrical installation to comply with the relevant requirements of an applicable British Standard, Harmonized or International Standard appropriate to the intended use of the equipment. It also offers alternative measures which may be followed where it is not possible to source equipment meeting British or Harmonized Standards, or where it is intended to use an item of equipment in a manner which is not covered by its standard.

133.2
133.3
133.4 Regulation group 133.2 requires equipment to be suitable in terms of voltage, current, frequency and power, while Regulations 133.3 and 133.4 respectively relate to conditions of installation and prevention of harmful effects on other equipment or the supply.

Section 314 deals with division of the installation with regard to factors such as facilitating the isolation of the installation or parts thereof, permitting safe inspection, testing, operation and maintenance, to enable the installation to be used safely.

341.1 Regulation 341.1 requires an assessment be made of the frequency and quality of maintenance that installed equipment, including switchgear, is likely to receive throughout the life of the installation.

422.2.2 Where switchgear is installed in the location of a fire evacuation escape route, Regulation 422.2.2 requires such switchgear to be installed so that it is accessible only to authorized persons (i.e. by lock and key). It also has to be installed in a non-combustible enclosure. Generally it would be better not to install any such switchgear or electrical equipment in any defined escape route as maintenance or other work on such switchgear etc may temporarily block or impede the escape route.

433.4 Section 433 deals with provision of devices for protection against overload currents, and Regulation 433.4 states that if a single protective device protects two or more conductors in parallel then no branch circuits may be installed and no devices providing isolation or switching may be placed in the parallel conductors.

434.1 Regulation 434.1 requires that the prospective fault current is to be determined at every relevant point in an installation. This requirement is directly related to the selection and specification of switchgear and protection and control devices.

Chap 46 Chapter 46 has been introduced into the 18th Edition of BS 7671 and provides the basic requirements for isolation and switching devices. The chapter should be read in conjunction with Section 537 of Chapter 53.

514.9.1
514.11.1
514.15.1 Amongst other information required to be provided within an installation by Regulation 514.9.1 are those details necessary to allow for the identification of each device provided to perform an isolation or switching function; this can be a chart for small (household) installations, but significantly more detail – such as "as built" drawings and operation and maintenance manuals – would be necessary in larger installations. Regulation 514.11.1 requires a notice to be posted wherever live parts are present which cannot be isolated by the operation of a single device. Regulation 514.15.1 requires notices to be posted where an installation contains a generator or additional sources of supply.

521.10.202 Regulation 521.10.202 requires that all cables routed in buildings must be fixed such that they cannot come away from their supports during fires and collapse to block escape routes or entangle firefighters searching the building in smoke filled conditions. Not all escape routes in a building are defined and signed – in open plan office areas there can be several possible routes away from a desk or meeting room across an open office area to an escape stair (see BS 9999) and all such routes should be kept clear from possible entanglement. An escape route is defined in the Wiring Regulations as being, "a path to follow for access to a safe area in the event of an emergency". The requirement for surface run wiring systems not to be subject to premature collapse in the event of a fire applies to all areas not just designated escape routes. Where cables run around or down walls – say to a socket-outlet – and are contained in small section plastic trunking the cables should be fixed inside the trunking if it would be possible for the trunking to deform in heat and the lid come off thus releasing the cables.

Sect 537 Specific requirements relating to isolation and switching are to be found in Section 537.

543.3.3.101 The continuity of earthing arrangements is fundamental to many of the measures for protection against electric shock, including the most commonly employed measure, automatic disconnection of supply (ADS). As a result it is not permitted to install a switching device in a protective conductor with the exception of:

▶ a changeover arrangement for installations supplied from more than one source of supply which are dependent upon separate earthing arrangements which cannot be connected at the same time

▶ multipole linked switching arrangements and plug-in devices so arranged that the protective conductor
- cannot be interrupted before the live conductors are disconnected
- is reconnected before or at the same time as the live conductors.

543.4.7 Regulation 543.4.7 prohibits the installation of an isolator or other switching device in the outer conductor of a concentric cable. (The ESQCR prohibits the installation of a TNC system supplied from the public supply network but old installations do still exist.)

At the completion of an installation guidance on the operation and use of the installation should be given in the documentation handed over to the client by the designer and installer on completion of the work. This will usually be contained in the Operation and Maintenance (O&M) manuals.

132.12
132.15.201
512.2
GN1
There is a general requirement that all electrical equipment, including isolating and switching devices, shall be so arranged to facilitate operation, inspection, fault detection, maintenance and repair. Such devices should also be suitable for the environment in which they are expected to be sited and remain operational, and in this regard reference should also be made to Guidance Note 1: *Selection & Erection*.

Regulation 15 of The Electricity at Work Regulations 1989 requires that in order to prevent injury, adequate working space, adequate means of access, and adequate lighting should be provided wherever electrical equipment is sited, so as to permit the safe use of such equipment in situations where danger may arise. The guidance given on Regulation 12(3) in HSR25 also expands on the need for items of switchgear to remain readily accessible.

The Construction (Design and Management) Regulations also make specific requirements regards access to equipment for installation, maintenance and removal/demolition.

133.1
511.1
GN3
Regulations 133.1 and 511.1 include the requirement that every item of equipment used in an electrical installation shall comply with the relevant requirements of the applicable British Standard or Harmonized Standard appropriate to the intended use of the equipment. Regulation 133.1.3 requires that if an item of equipment used is not in accordance with such standards or is used outside the scope of its standard the designer must confirm that the equipment provides at least the same degree of safety as that provided by compliance with the regulations and this is to be noted on the appropriate electrical installation certification (see Guidance Note 3). Where there are no applicable standards to cover certain equipment the use of such equipment must be agreed between the designer, installer and the client.

511.2
If equipment is to be used that complies with a foreign national standard based on an IEC standard the designer or other person responsible for specifying the installation must verify that any differences between that standard and the corresponding British or Harmonized Standard will not result in a lesser degree of safety than that afforded by the British or Harmonized Standard and its use noted on the appropriate electrical installation certification (see IET Guidance Note 3 for further information).

Every item of equipment must be suitable for the nominal voltage of the installation or part of the installation taking into account the highest and/or lowest voltage that may occur in normal service. The use of such equipment must be noted on the appropriate electrical installation certification.

The devices must also be suitable for switching duty to break and make the circuit load current and break the circuit fault current where required.

- A circuit-breaker by its function can break load current, so is suitable for **on-load** isolation, i.e. disconnection whilst carrying load current.
- A disconnector or isolator is suitable for **off-load** isolation. (Note the definitions in Part 2 of BS 7671).
- A switch-disconnector or isolating switch is suitable for **on-load** isolation.

If a device has "switch" in its description, it's suitable for on-load; if it doesn't, it's off-load.

The product standards for household and similar generally use the term "isolating switch" e.g. BS EN 60669-2-4 for a shower isolating switch.

The product standards for commercial/industrial applications, for example, BS EN 60947-3 generally use the terms "switch-disconnector", "switch-fuses", "fuse-switches" etc.

Electrical equipment standards

It is not practical to list all the relevant electrical equipment standards in any guidance document. Standards that are mentioned in BS 7671 are listed in Appendix 1 of that document for guidance – however there are many more! Some major standards relevant to isolation and switching are considered below.

BS EN 60073:2002 *Basic and safety principles for man-machine interface, marking and identification. Coding principles for indicators and actuators* provides details and rules for colours, shapes and positions of indicating devices and actuators.

▶ BS EN 60947 *Low-voltage switchgear and controlgear* has multiple parts which cover specific products. These parts contain a number of requirements specific to the isolation function on factors such as performance, indication and marking.
▶ BS EN 60669-1:1999+A2:2008 *Switches for household and similar fixed electrical installations. General requirements* (current but being revised)
For example: mechanically operated and intended for functional purposes only.

▶ BS EN 60669-2-1 / -2 / -3 / Switches for household and similar fixed electrical installations intended for functional purposes only covering:
- 2-1 Electronic switches
- 2-2 Electromagnetic remote-control switches (RCS)
- 2-3 Time delay switches (TDS)

▶ BS EN 60669-2-4 Particular requirements for Isolating switches for household and similar fixed electrical installations

BS EN 60669-2-6:2012 *Switches for household and similar fixed electrical installations. Particular requirements. Fireman's switches for exterior and interior signs and luminaires.*

▶ BS 1363-2:2016 *13 A plugs, socket-outlets, adaptors and connection units.* (BS 1363-2:1995+A4:2012, remains current however and will be withdrawn on 31 August 2019).
▶ BS EN 60309-1:1999+A2:2012 *Plugs, socket-outlets and couplers for industrial purposes. General requirements.*

There are several systems, such as electronic lighting control by building management systems, which may be made up from many units or components that have no specific product standard. In such cases, the individual units or components should be manufactured and installed to relevant standards and the installation should comply with BS 7671. For home and building electronic systems (HBES) see also BS EN 60669-2-5:2016 (BS EN 50428:2005+A2:2009 remains current however and will be withdrawn on 31 August 2020).

Table 537.4 Table 537.4 in BS 7671 gives guidance on the selection of appropriate devices to perform protective, isolation and switching functions and is reproduced here as Table 3.1. It must be noted that Table 3.1 can only provide guidance and it is for the designer to consider the necessary requirements for their specific application and whether a specific device is suitable for the required application. Also in Table 3.1, the functions provided by the devices for isolation and switching are summarized, together with an indication of the relevant product standards.

▼ **Table 3.1** Guidance on the selection of protective, isolation and switching devices

Device	Standard	Isolation[4]	Emergency switching[2]	Functional switching
Switching device	BS EN 50428	No	No	Yes
	BS EN 60669-1	No	Yes	Yes
	BS EN 60669-2-1	No	No	Yes
	BS EN 60669-2-2	No	Yes	Yes
	BS EN 60669-2-3	No	Yes	Yes
	BS EN 60669-2-4	Yes[3]	Yes	Yes
	BS EN 60947-3	Yes[1,3]	Yes	Yes
	BS EN 60947-5-1	No	Yes	Yes
Contactor	BS EN 60947-4-1	Yes[1,3]	Yes	Yes
	BS EN 61095	No	No	Yes
Circuit-breaker	BS EN 60898	Yes[3]	Yes	Yes[5]
	BS EN 60947-2	Yes[1,3]	Yes	Yes[5]
	BS EN 61009-1	Yes[3]	Yes	Yes[5]
RCD	BS EN 60947-2	Yes[1,3]	Yes	Yes[5]
	BS EN 61008-1	Yes[3]	Yes	Yes[5]
	BS EN 61009-1	Yes[3]	Yes	Yes[5]
Isolating switch	BS EN 60669-2-4	Yes[3]	Yes	Yes
	BS EN 60947-3	Yes[1,3]	Yes	Yes
Plug and socket-outlet (≤ 32 A)	BS EN 60309	Yes[3]	No	Yes
Plug and socket-outlet (> 32 A)	BS EN 60309	Yes[3]	No	No
Device for connection of luminaire	BS EN 61995-1	Yes[3]	No	No
Control and protective switching device for equipment (CPS)	BS EN 60947-6-1	Yes[1,3]	Yes	Yes
	BS EN 60947-6-2	Yes[1,3]	Yes	Yes
Fuse	BS 88 series	Yes	No	No

▼ **Table 3.1** *Continued*

Device	Standard	Isolation[4]	Emergency switching[2]	Functional switching
Device with semiconductors	BS EN 50428	No	No	Yes
	BS EN 60669-2-1	No	No	Yes
Luminaire supporting coupler	BS 6972	Yes[3]	No	No
Plug and unswitched socket-outlet	BS 1363-1	Yes[3]	No	Yes
	BS 1363-2	Yes[3]	No	Yes
Plug and switched socket-outlet	BS 1363-1	Yes[3]	No	Yes
	BS 1363-2	Yes[3]	No	Yes
Plug and socket-outlet	BS 5733	Yes[3]	No	Yes
Switched fused connection unit	BS 1363-4	Yes[3]	Yes	Yes
Unswitched fused connection unit	BS 1363-4	Yes[3] (removal of fuse link)	No	No
Fuse	BS 1362	Yes	No	No
Cooker control unit switch	BS 4177	Yes[3]	Yes	Yes

Notes to Table 3.1:
Yes = Function provided; No = Function not provided

1 Function provided if the device is suitable and marked with the symbol for isolation. (See IEC 60617 identity number S00288.)

2 See Regulation 537.3.3.6.
3 Device is suitable for on-load isolation,i.e. disconnection whilst carrying load current.
4 In an installation forming part of a TT or IT system, isolation requires disconnection of all the live conductors. See Regulation 462.2.
5 Circuit breakers and RCDs are primarily circuit-protective devices and, as such, they are not intended for frequent load switching. Infrequent switching of circuit-breakers on-load is admissible for the purposes of isolation or emergency switching.
For a more frequent duty, the number of operations and load characteristics according to the manufacturer's instructions should be taken into account or an alternative device from those listed as suitable for functional switching in Table 537.4 should be employed.
6 It is recognised that whilst a plug and socket may not be selected for an emergency switch (Regulation 537.3.3.3, they maybe used in practice for this purpose

Note: An entry of (1,3 in the notes above) means that the device is suitable for on-load isolation only if it is marked with the symbol for on-load isolation.

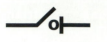

3.2   Isolation

The definition given in BS 7671 is:

Part 2 *A function intended to make dead for reasons of safety all, or a discrete section, of the electrical installation by separating the electrical installation or section thereof, from every source of electrical energy.*

Regulation 12(1) of The Electricity at Work Regulations 1989 requires that:

> **(1)** Where necessary to prevent danger, suitable means (including, where appropriate, methods of identifying circuits) shall be available for –
> **(a)** cutting off the supply of electrical energy to any electrical equipment; and
> **(b)** the isolation of any electrical equipment.

Regulation 12(2) defines isolation as:

> *... the disconnection and separation of the electrical equipment from every source of electrical energy in such a way that this disconnection and separation is secure.*

The Electricity at Work Regulations 1989. Guidance on Regulations HSR25 published by the Health and Safety Executive advises, with reference to Regulation 12(1)(b) above, that isolation is the process of ensuring that the supply to all or a particular part of an installation remains switched off and that inadvertent reconnection is prevented.

The issue of preventing inadvertent reconnection is covered in Regulation 13:

> **13. Precautions for work on equipment made dead**
>
> Adequate precautions shall be taken to prevent electrical equipment, which has been made dead in order to prevent danger while work is being carried out on or near that equipment, from becoming electrically charged during that work if danger may thereby arise.

The 'security' mentioned in Regulation 12(2) may be achieved either as a result of the means of isolation remaining directly under the control of the persons who are reliant upon its remaining effective or, where this is not the case, by the application of a locking device being used to secure the means of isolation in the OFF position. In practice this will nearly always result in the use of padlocks being used.

462.3 Neither The Electricity at Work Regulations 1989 nor BS 7671 calls for the application
537.2.4 of a means of locking to be applied in all cases, but Regulation 462.3 identifies
537.2.5 locking as one means of preventing inadvertent closure of the isolator. However, if any possibility exists of the means of isolation being compromised, some form of locking should be applied. Indeed, Regulation 537.2.4 requires the means of isolation to be securable by placing it in a lockable space or lockable enclosure or by padlocking or other suitable means. In reality the isolation device is usually secured by a quality "safety" padlock for which there is only ONE key available, and this is retained by the

operative managing the isolation. Lockable spaces or enclosures may be insecure or have multiple keys available and access control cannot be guaranteed (see Appendix D for further guidance on safety padlocks).

The complexity of isolation and switching arrangements increases with the size of the electrical installation. Figure 3.1 shows switchboards suitable for use in large industrial or commercial installations.

▼ **Figure 3.1** Switchboards suitable for use in industrial or commercial installations awaiting delivery to site [photograph courtesy of Schneider]

3.3   Switching off for mechanical maintenance

The term 'mechanical maintenance' is defined in BS 7671 as:

Part 2   *The replacement, refurbishment or cleaning of lamps and non-electrical parts of equipment, plant and machinery.*

Broadly speaking, switching off for mechanical maintenance is a function similar to isolation whereby electrically actuated equipment is made safe for persons to work on, in, or near the equipment. It differs from the requirement for isolation as it is not intended to remove terminal covers or provide access to any electrical parts. It is used to switch off machines to perhaps allow operators or persons who are not electrically skilled to safely access parts of a machine for cleaning or to make non electrical adjustments or repairs.

It is unusual in the UK to install a special switching device in the electrical supply to a machine specifically for mechanical maintenance use, and usually the isolator is used for both isolation and switching off for mechanical maintenance.

HSR25   Although there is no reference to this operation in The Electricity at Work Regulations, the guidance given on Regulation 12(1)(a) in the *The Electricity at Work Regulations 1989. Guidance on Regulations* HSR25 acknowledges that 'there may be a need to switch off electrical equipment for reasons other than preventing electrical danger but these considerations are outside the scope of the [Electricity at Work] Regulations' or other safety regulations where there is no 'danger' in the work.

131.1(v) It should be remembered that the scope of BS 7671 includes the protection of persons from risk of injury from mechanical movement of electrically actuated equipment, by the employment of electrical emergency switching or electrical switching for mechanical maintenance of non-electrical parts of such equipment.

The purpose of the measure is to prevent physical injury, but not electric shock or burns, as mechanical maintenance should not involve work upon, access to, or exposure of normally live parts. Particular attention is drawn therefore to the requirements of Regulations 14 and 16 of the EWR 1989 in respect of working on or near live parts.

Switching off for mechanical maintenance should be considered if access to machinery or equipment may involve access to normally moving parts. Isolation of supplies to machinery or equipment may be more appropriate in some situations to provide a sufficient degree of physical safety.

3.4 Emergency switching

BS 7671 defines this as:

Part 2 *An operation intended to remove, as quickly as possible, danger, which may have occurred unexpectedly.*

465.1 Regulation 465.1 requires that means are to be provided for emergency switching off of any part of an installation where it may be necessary to control the supply to remove an unexpected danger, however Regulation 465.4 requires that the arrangement of the emergency switching shall be such that its operation does not introduce a further danger or interfere with the complete operation necessary to remove the danger, which requires a considerable risk assessment and very good understanding of the operation of the machinery system involved. Regulation 465.4 then also has a note to point out that the operation of the switching device is to be understood as switching off in case of emergency and switching on to reenergise the relevant circuit.

537.3.3 It is recognised in the definition that this danger may arise from non-electrical events or occurrences, however regulation Sections 465 and 537.3.3 only recognise "Emergency Switching Off" to switch off the electrical supply where a risk of electric shock or another electrical risk is involved. In previous editions of BS 7671 it was considered that 'Emergency switching may be emergency switching on or emergency switching off' as just removing a supply may allow other events to happen, for example, the release of an electromagnet brake.

Section 537 requires that devices for emergency switching off are to provide switching of the main electrical circuit rather than local control circuits, and where electrically powered equipment is within the scope of BS EN 60204, the requirements for emergency switching off of that standard apply. The means for emergency switching off may consist of either one switching device capable of directly cutting off the appropriate supply, or a combination of devices activated by a single action for the purpose of cutting off the appropriate supply.

It must be noted that plugs and socket-outlets are not to be selected or provided for use as means for emergency switching off.

3.5 Emergency stopping

The definition given in BS 7671 is:

Part 2   *Emergency switching intended to stop an operation.*

537.3.3.3   Again, it should be appreciated that the operation of an emergency stopping device may, in some cases, need to allow the continued supply to electrically actuated brakes and as such not remove ALL sources of supply. This is obviously an important factor when considering suitable means for achieving safe isolation. HSR25 reminds readers of this in its discussion of Regulation 12(1)(b), where it states that 'it must be understood that the two functions of switching off (which includes emergency stopping) and isolation are not the same, even though in some circumstances they are performed by the same action or by the same equipment'.

Regulation 16 of The Provision and Use of Work Equipment Regulations 1998 provides specific requirements for machine emergency stop controls and reference should also be made to the following standards:

- ▶ BS EN ISO 13850:2008 *Safety of machinery. Emergency stop. Principles for design*
- ▶ BS EN 60204-1:2006 + A1:2009 *Safety of machinery. Electrical equipment of machines. General requirements*
- ▶ BS EN 60947-5-5:1998+A11:2013 *Low-voltage switchgear and controlgear. Control circuit devices and switching elements. Electrical emergency stop devices with mechanical latching function.*

Generally an electrical installer will only be concerned with the provision of electrical supplies and control wiring to machines and machine operating systems designed and provided by specialist manufacturers. The design of emergency and safety systems requires detailed experience and a complete understanding of the equipment operation and legislative requirements.

3.6 Functional switching

This is defined in BS 7671 as:

Part 2   *An operation intended to switch ON or OFF or vary the supply of electrical energy to all or part of an installation for normal operating purposes.*

Sect 463   As BS 7671 is mainly concerned with the safety of electrical installations it only deals
537.3.1   with the operation of functional switching and functional switching devices generally.
Appendix 17   There are many possible variations for functional control, and generally lighting is the major visible subject for such control, but followed closely by heating and ventilation. Although BS 7671 generally does not regulate for functional controls they are necessary to aid in energy conservation and are mentioned in the new Appendix 17 in BS 7671. There are also specific requirements for controls for energy conservation in the UK Building Regulations.

Table 537.4   BS 7671 does also recognize that some functional switching or control devices may also be employed to provide variously the functions of isolation, emergency switching, or switching off for mechanical maintenance purposes. However, it is incumbent upon the designer to confirm that any device selected to provide more than one switching function in an installation meets the relevant requirements of BS 7671 for each such function. Table 537.4 of BS 7671 provides guidance on the selection of protective, isolation and switching devices.

Functional switching must be provided for each part of a circuit which may require to be controlled independently of other parts of the installation. Functional switching devices are to be suitable for the most onerous duty they are intended to perform, and the characteristic of the load to be switched shall be considered (e.g. utilization category) – see Table 3.2 for further details. In general, all current-using equipment requiring control must be controlled by an appropriate functional switching device, and a single functional switching device may control several items of current -using equipment intended to operate simultaneously.

3.7 Ordinary, instructed and skilled persons

These are defined in BS 7671.

Part 2 **Ordinary person.** *Person who is neither a skilled person nor an instructed person.* This would include householders and other persons such as teachers and office workers who utilise electrical installations and equipment in their normal work but who have no specialist electrical understanding.

Instructed person (electrically). *Person adequately advised or supervised by a skilled person (as defined) to enable that person to perceive risks and to avoid hazards which electricity can create.*

Skilled person (electrically). *Person who possesses, as appropriate to the nature of the electrical work to be undertaken, adequate education, training and practical skills, to perceive risks and to avoid hazards which electricity can create.*

Note: The term "electrically" is presumed to be present whenever either a skilled or instructed person or persons are mentioned in BS 7671.

It must be remembered that advice on its own will not turn an ordinary person into an instructed person, and it is always the responsibility of the skilled person to ensure themselves that the instructed person has understood the instructions. This can be done by observing the instructed person carrying out their training in a test situation.

Regulation 16 of The Electricity at Work Regulations 1989 requires persons to be competent to prevent danger and injury, and it is for an employer to assure themselves that any skilled, instructed or unskilled ordinary persons that they require to undertake any work are competent to do so. There is no specific definition of competence but HSE publication HSR25 provides guidance.

The term "Authorised Person" is not used in BS 7671 but is widely used in industry to denote a skilled person who has had special training in switching and is experienced in operating a specific complex installation or installations and has been specifically appointed as a specialist to carry out switching in those installations.

In this Guidance Note any person referred to is presumed to be a skilled and competent person unless noted otherwise. The use of the word "person" in the singular can be assumed to include the plural also where more than one person is required. It is for an employer or contractor to provide adequate persons to carry out work safely. Generally when switching anything but the smallest system an authorised person will have a second competent person with them for safety.

3.8 Definitions from other standards

Table 537.4 Table 537.4 lists the product standards appropriate to the commonly available isolation and switching devices. These standards contain further definitions taken directly from, or based upon, those given in the International Electrotechnical Vocabulary (IEV 60050), published by the International Electrotechnical Commission (IEC), a number of which are reproduced below.

(a) **Breaking capacity** (of a switching device or fuse)

Value of prospective breaking current that a switching device or a fuse is capable of breaking at a stated voltage under prescribed conditions of use and behaviour.

Note: The voltage to be stated and the conditions to be prescribed are dealt with in the relevant product standard.

For AC, the current is expressed as the symmetrical rms value of the AC component.

Note: For switching devices, the breaking capacity may be termed according to the kind of current included in the prescribed conditions, e.g. line-charging breaking capacity, cable charging breaking capacity, single capacitor bank breaking capacity, etc.

For *short-circuit breaking capacity*, see **(h)**.

(b) **Breaking current** (of a switching device or fuse)

The current in a pole of a switching device or in a fuse at the instant of initiation of the arc during a breaking process.

Note: For AC, the current is expressed as the symmetrical rms value of the AC component.

(c) **Conditional short-circuit current** (of a circuit or switching device)

Prospective current that a circuit or a switching device, protected by a specified short-circuit protective device, can satisfactorily withstand for the total operating time of that device under specified conditions of use and behaviour.

Note: For the purpose of this standard, the short-circuit protective device is generally a circuit-breaker or a fuse.

Note: This definition differs from IEV 441-17-20 by broadening the concept of current-limiting device into a short-circuit protective device, the function of which is not only to limit the current.

(d) **Fuse-combination unit**

Combination of a mechanical switching device and one or more fuses in a composite unit, assembled by the manufacturer or in accordance with his instructions.

(e) **Isolator**

A mechanical switching device which, in the open position, complies with the requirements specified for the isolating function. An isolator is otherwise known as a disconnector.

(f)   Isolation

Function intended to make dead for reasons of safety all, or a discrete section, of the electrical installation by separating the electrical installation, or section thereof, from every source of electrical energy.

(g)   Making capacity (of a switching device)

Value of prospective making current that a switching device is capable of making at a stated voltage under prescribed conditions of use and behaviour.

Note: The voltage to be stated and the conditions to be prescribed are dealt with in the relevant specifications.

(h)   Rated value

A quantity value assigned, generally by the manufacturer, for a specified operating condition of a component, device or equipment.

Note: Examples of rated value usually stated for fuses: voltage, current, breaking capacity.

(i)   Rating

The set of rated values and operating conditions.

(j)   Short-circuit breaking capacity

Breaking capacity for which prescribed conditions include a short-circuit at the terminals of the switching device.

(k)   Short-circuit making capacity

Making capacity for which the prescribed conditions include a short-circuit at the terminals of the switching device.

(l)   Switch-disconnector

Switch which, in the open position, satisfies the isolating requirements specified for a disconnector.

(m)   Disconnector

A mechanical switching device which provides, in the open position, an isolating distance in accordance with specified requirements.

Note: A disconnector is capable of opening and closing a circuit when either negligible current is broken or made, or when no significant change in the voltage across the terminals of each of the poles of the disconnector occurs. It is also capable of carrying currents under normal circuit conditions and carrying for a specified time currents under abnormal conditions such as those of short circuit.

(n)   Utilization category (for a switching device or a fuse)

A combination of specified requirements related to the condition in which the switching device or the fuse fulfils its purpose, selected to represent a characteristic group of practical applications.

Note: The specified requirements may concern, for example, the values of making capacities (if applicable), breaking capacities and other characteristics, the associated circuits and the relevant conditions of use and behaviour.

The utilization category of equipment defines the intended application and is specified in the relevant product standard and is characterised by one or more of the following service conditions:

▶ current(s), expressed as multiple(s) of the rated operational current
▶ voltage(s), expressed as multiple(s) of the rated operational voltage
▶ power factor or time constant
▶ short-circuit performance
▶ selectivity
▶ other service conditions, as applicable.

Table 3.2 provides a comprehensive list of utilization categories for low voltage switchgear and controlgear and is based on BS EN 60947-1:2007+A2:2011 and BS EN 61095:2009.

▼ **Table 3.2** Examples of utilization categories for low voltage switchgear and controlgear

Nature of current	Category	Typical applications	Relevant BS EN product standard
AC	AC-20	Connecting and disconnecting under no-load conditions	60947-3
	AC-21	Switching of resistive loads, including moderate overloads	
	AC-22	Switching of mixed resistive and inductive loads, including moderate overloads	
	AC-23	Switching of motor loads or other highly inductive loads	
AC	AC-1	Non-inductive or slightly inductive loads, resistance furnaces	60947-4-1
	AC-2	Slip-ring motors: starting, switching off	
	AC-3	Squirrel-cage motors: starting, switching off motors during running	
	AC-4	Squirrel-cage motors: starting, plugging,[a] inching[b]	
	AC-5a	Switching of electric discharge lamp controls	
	AC-5b	Switching of incandescent lamps	
	AC-6a	Switching of transformers	
	AC-6b	Switching of capacitor banks	
	AC-7a	Slightly inductive loads	
	AC-7b	Motor Loads	
	AC-8a	Hermetic refrigerant compressor motor control with manual resetting of overload releases	

Nature of current	Category	Typical applications	Relevant BS EN product standard
	AC-8b	Hermetic refrigerant compressor motor control with automatic resetting of overload releases	
AC	AC-7c	Switching of compensated electric discharge lamp controlᶜ	61095
AC	AC-52a	Control of slip-ring motor stators: 8 h duty with on-load currents for start, acceleration, run	60947-4-2
	AC-52b	Control of slip-ring motor stators: intermittent duty	
	AC-53a	Control of squirrel-cage motors: 8 h duty with on-load currents for start, acceleration, run	
	AC-53b	Control of squirrel-cage motors: intermittent duty	
	AC-58a	Control of hermetic refrigerant compressor motors with automatic resetting of overload releases: 8 h duty with on-load currents for start, acceleration, run	
	AC-58b	Control of hermetic refrigerant compressor motors with automatic resetting of overload releases: intermittent duty	
AC	AC-51	Non-inductive or slightly inductive loads, resistance furnaces	60947-4-3
	AC-55a	Switching of electric discharge lamp controls	
	AC-55b	Switching of incandescent lamps	
	AC-56a	Switching of transformers	
	AC-56b	Switching of capacitor banks	
AC	AC-12	Control of resistive loads and solid-state loads with isolation by optocouplers	60947-5-1
	AC-13	Control of solid-state loads with transformer isolation	
	AC-14	Control of small electromagnetic loads	
	AC-15	Control of AC electromagnetic loads	
AC	AC-12	Control of resistive loads and solid-state loads with optical isolation	60947-5-2
	AC-140	Control of small electromagnetic loads with holding (closed) current ≤ 0.2 A, e.g. contactor relays	
AC	AC-31	Non-inductive or slightly inductive loads	60947-6-1
	AC-33	Motor loads or mixed loads including motors, resistive loads and up to 30 % incandescent lamp loads	

Nature of current	Category	Typical applications	Relevant BS EN product standard
	AC-35	Electric discharge lamp loads	
	AC-36	Incandescent lamp loads	
AC	AC-40	Distribution circuits comprising mixed resistive and reactive loads having a resultant inductive reactance	60947-6-2
	AC-41	Non-inductive or slightly inductive loads, resistance furnaces	
	AC-42	Slip-ring motors; starting, switching off	
	AC-43	Squirrel-cage motors: starting, switching off motors during running	
	AC-44	Squirrel-cage motors: starting, plugging,[a] inching[b]	
	AC-45a	Switching of electric discharge lamp controls	
	AC-45b	Switching of incandescent lamps	
AC and DC	A	Protection of circuits, with no rated short-time withstand current	60947-2
	B	Protection of circuits, with a rated short-time withstand current	
DC	DC-20	Connecting and disconnecting under no-load conditions	60947-3
	DC-21	Switching of resistive loads, including moderate overloads	
	DC-22	Switching of mixed resistive and inductive loads, including moderate overloads (e.g. shunt motors)	
	DC-23	Switching of highly inductive loads (e.g. series motors)	
DC	DC-1	Non-inductive or slightly inductive loads, resistance furnaces	60947-4-1
	DC-3	Shunt-motors, starting, plugging,[a] inching.[b] Dynamic braking of motors	
	DC-5	Series-motors, starting, plugging,[a] inching.[b] Dynamic braking of motors	
	DC-6	Switching of incandescent lamps	
DC	DC-12	Control of resistive loads and solid-state loads with isolation by optocouplers	60947-5-1
	DC-13	Control of electromagnets	
	DC-14	Control of electromagnetic loads having economy resistors in circuit	
DC	DC-12	Control of resistive loads and solid-state loads with optical isolation	60947-5-2
	DC-13	Control of electromagnets	

Isolation

4

4.1 General

132.10
Chap 46
Sect 537

The fundamental principle for isolation is that effective means suitably placed shall be provided so that all voltage may be cut off from every installation, from every circuit thereof and from all equipment, as may be necessary to prevent or remove danger. As a result of the restructuring within the 18th Edition, the requirements relating to isolation are now to be found in both Chapter 46 and in Section 537 of Chapter 53 – the basic requirements being in Chapter 46 and the application of these requirements being in Section 537.

422.3.13 In TT and IT systems serving locations with risks of fire due to the nature of processed or stored materials, such as woodworking shops, industrial scale bakeries, paper mills and the like, every circuit should be provided with a means of isolation which disconnects all live conductors by means of a linked switch or linked circuit-breaker.

Fuses and associated neutral links may be used to facilitate isolation. However, the use of fuses and links should be restricted to electrically skilled or instructed persons to avoid any risks associated with their being withdrawn or replaced on load.

In general, whichever form of isolation is used (fuse withdrawal, circuit-breaker operation, etc.) a caution notice, in accordance with the Health and Safety (Safety Signs and Signals) Regulations 1996 (see ACOP L64 for guidance), must be applied at or near the point of isolation. See Figure 4.1.

A particular danger may arise in older single-phase installations where double-pole fuseboards may still be found. To avoid any confusion it is recommended that both fuses protecting a circuit are removed. In any case, isolation should be proved by the use of an approved voltage detector. Remember to check the operation of the voltage detector before and after use.

462.1
514.11.1
537.1.2

Where an installation, item of equipment or enclosure contains live parts which require the operation of more than one isolating device in order to disconnect all sources of supply, a notice warning of the need to disconnect more than one supply should be posted such that it will be seen by a person wishing to gain access to said installation, item of equipment or enclosure. See Figure 4.1.

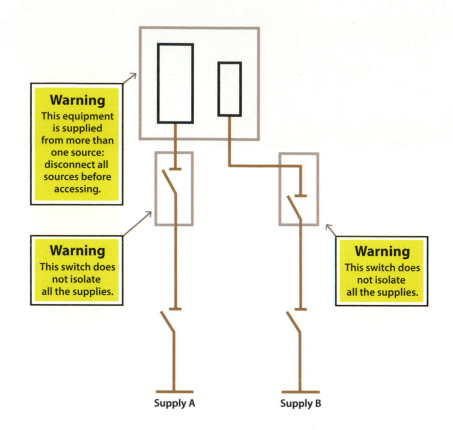

▼ **Figure 4.1** Examples of labelling where one switch does not isolate equipment

Warning
This equipment is supplied from more than one source: disconnect all sources before accessing.

Warning
This switch does not isolate all the supplies.

Warning
This switch does not isolate all the supplies.

Supply A   Supply B

537.1.2 It has been frequently questioned whether the requirement for a warning notice in accordance with Regulation 537.1.2 applies to a two-gang (or larger) light switch fed by two (or more) separate circuits in a domestic situation. It has always been impractical to require a visible warning notice in a domestic location as it could blight the householders decorations and will be removed, but it is suggested that a warning label is placed just under the front plate of a "grid-type" switch or visibly in the box of a plate switch.

4.2   Isolation at the origin of an installation

462.1.201 Each installation must be provided with a means of disconnecting from the supply at the origin of the installation.

537.1.3 The distributor's switchgear may be used as a means for isolation where a linked disconnector or circuit breaker has been provided.

This means of isolation should be placed as near as practicable to the origin of the installation and should be suitable to allow for switching the supply on load. In single-phase supplies in domestic and similar properties to be used by ordinary persons (as defined in BS 7671) this switchgear should be double-pole so as to break both the live conductors. See Figure 4.2(a), (b) and (e).

Devices suitable for isolation are to be selected according to the requirements which are based on the overvoltage categories applicable at their point of installation and shall be designed for over voltage category III or IV except the plug of a plug and socket-outlet combination identified in Table 3.1 as suitable for isolation. Where electrically powered equipment is within the scope of BS EN 60204, the requirements for isolation of that standard are to be apply.

461.2 In general, the neutral conductor of a supply derived from a TN-S or PME system
Sect 537 in accordance with the Electricity Safety, Quality and Continuity Regulations can be
537.1.2 considered to be reliably connected to Earth by a suitably low impedance. Where this
is the case, it is not necessary to isolate or switch the neutral conductor except as
mentioned above – see Figures 4.2 and 4.3. Combined protective and neutral (PEN)
conductors are not commonplace in installations within the UK and are not permitted
in installations supplied from the public supply network. Where they are installed, they
should not be isolated or switched.

▼ **Figure 4.2** Isolation at the origin of typical single-phase installations

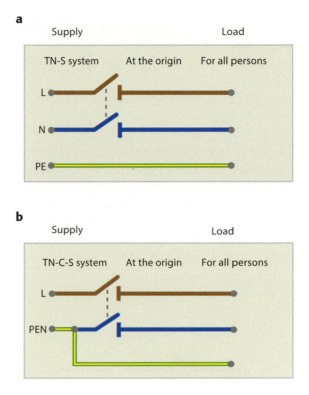

See Regulation 514.4.3 – A PEN conductor shall, when insulated, be marked by one
of the following methods:
  (i) Green-and-yellow throughout its length with, in addition, blue markings at
  the terminations
  (ii) Blue throughout its length with, in addition, green-and-yellow markings at
  the terminations.

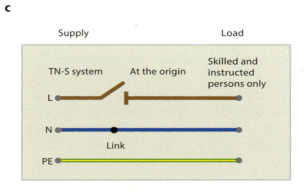

d

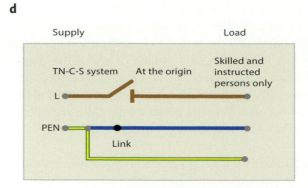

See Regulation 514.4.3 – A PEN conductor shall, when insulated, be marked by one of the following methods:

(i) Green-and-yellow throughout its length with, in addition, blue markings at the terminations

(ii) Blue throughout its length with green-and-yellow markings at the terminations.

462.1
537.1.2
537.1.5
Where an installation is supplied from more than one source, which is increasingly the case even for domestic and other small premises, and these sources are required to have their own independent earthing arrangements, a changeover switching arrangement should be provided between the neutral point and means of earthing to ensure that not more than one means of earthing is effective at any one time. This changeover of the earth connection should occur at substantially the same time as the changeover of the associated live conductors.

e

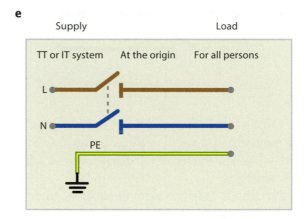

A main switch is required for each source of supply to an installation. These main switches may be arranged such that they can be operated at the same time, via a suitable interlock arrangement, to provide disconnection of the installation from all sources of supply. Where this is not the case, a warning notice will be required informing of the need to operate all main switches to fully isolate the installation.

537.2.7 This situation frequently occurs when an additional consumer unit is added, for example, to supply an electric shower. Regulation 537.2.7 requires each device used for isolation to be clearly identified by position or durable marking to indicate the installation it isolates.

▼ **Figure 4.3** Isolation at the origin of a three-phase installation

a

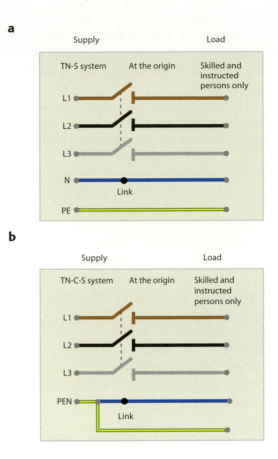

b

See Regulation 514.4.3 – A PEN conductor shall, when insulated, be marked by one of the following methods:

(i) Green-and-yellow throughout its length with, in addition, blue markings at the terminations

(ii) Blue throughout its length with green-and-yellow markings at the terminations.

c

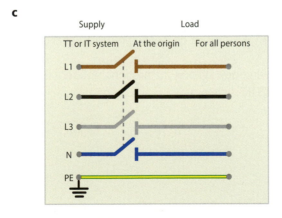

4.3   Isolation of circuits within an installation

462.2   Provision should be made such that each circuit within an installation can be isolated from its source(s) of supply. As was the case with reference to 462.3 (discussed previously) for main isolation, in an installation being part of a TN-S or TN-C-S system, this will not require the neutral to be switched. See Figure 4.4 (a) and (c).

It is permitted to isolate more than one circuit by a single device. This might be appropriate, for example, in the case of a production line consisting of a number of pieces of equipment controlled from a single control panel.

537.2.4   Regulation 537.2.4 requires suitable means to be provided to prevent any item of
537.2.5   equipment from becoming inadvertently or unintentionally energized. It is generally assumed that this will be achieved by either locking the means of isolation in the OFF position, or on account of the device remaining under the direct control of the persons reliant upon the isolation, remaining effective.

462.4   On occasion, equipment may contain components which retain a charge for some time after the supply has been disconnected. Where this is the case, a means should be provided to effect their discharge.

537.2.7   It is quite acceptable for an isolator to be placed remotely from the equipment it isolates. However, where this is the case, the means of isolation must be capable of being secured in the OFF or open position. The isolator should be labelled to indicate which equipment it isolates.

537.2.8   Although it is not a requirement to isolate or switch the neutral conductor in installations supplied from a TN supply system, it is still necessary to provide a means for disconnecting the neutral conductor. It should only be possible to disconnect the neutral conductor by the use of a tool, **not** a coin or similar.

560.7.5   Switchgear and controlgear provided in connection with a safety service should be installed such that it is accessible only to skilled or instructed persons (as defined).

Reference should also be made to the relevant standard for the safety service in question. In the case of fire alarm systems in industrial and commercial premises, reference should also be made to the requirements of BS 5839-1:2017 and for fire detection and alarm systems in domestic premises, section 15 of BS 5839-6:2013.

BS 5266-1   In the case of emergency lighting installations, reference should be made to clause 8.3 of BS 5266-1:2016.

▼ **Figure 4.4** Examples showing isolation arrangements for circuits within an installation

Single-phase

a

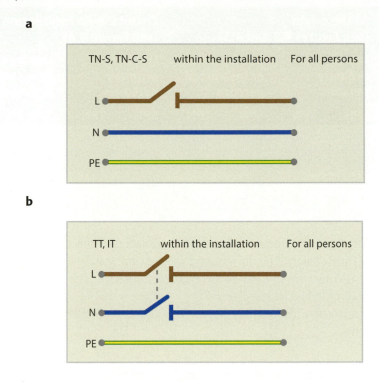

b

Three-phase

c

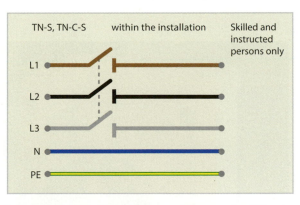

Note: ordinary persons would not normally be operating TPN systems.

d

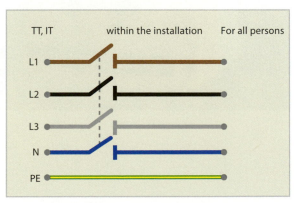

▼ **Table 4.1** Isolation requirements of the neutral conductor – see Figures 4.2, 4.3 and 4.4 previously.

| | At the origin (see Figures 4.2 and 4.3) | | | | Circuits in an installation (see Figure 4.4) | | | |
| | Use by ordinary persons | | Use by skilled or instructed persons | | Use by ordinary persons | | Use by skilled or instructed persons | |
	SPN	TPN	SPN	TPN	SPN	TPN	SPN	TPN
TN-S	YES	N/A	YES*	NO	NO	N/A	NO	NO
TN-C-S	YES	N/A	YES*	NO	NO	N/A	NO	NO
TT	YES	N/A	YES*	YES	YES	N/A	YES	YES
IT	YES	N/A	YES*	YES	YES	N/A	YES	YES

* Regulation 462.1.201 requires Double Pole isolation at the origin in a residential (dwelling) installation. This is required in such an installation regardless of the intended use by ordinary, skilled or instructed persons, but it is not required in commercial or industrial installations

N/A – Not applicable as ordinary persons would not normally be operating TPN systems

4.4   Devices for isolation and securing isolation

▼ **Figure 4.5** Switch-disconnector capable of being locked in the OFF position [photograph courtesy of Hager]

▼ **Figure 4.6** Types of lock-out designed for use with a padlock to secure circuit-breakers, RCBOs and RCDS

Apart from installations supplied from a TN supply system in accordance with the Electricity Safety, Quality and Continuity Regulations where disconnection of the neutral conductor is not required, Regulation 461.2 requires the provision of a device to isolate all live supply conductors from the circuit concerned. See Figure 4.4(b) and (d).

The device so selected must be suitable for the overvoltage category appropriate at its point of installation.

This regulation also makes it clear that semiconductor devices may not be used as isolating devices as they do not provide any physical break in the conductor, i.e. they cannot meet the dimensional contact separation requirements for an isolating device. They may, however, be used to initiate the operation of a main isolator.

supply system. Note that a generator in terms of Section 551 is any source of voltage generator alternative to the public supply such as photovoltaic, batteries and similar. Ways of achieving this requirement include the following examples:

▶ an interlock arrangement between the operating mechanisms or control circuits of the changeover switches – the interlock may be electrical, mechanical or electromechanical in operation
▶ a system of locks having a single transferable key
▶ a three position break-before-make changeover switch
▶ an automatic changeover switch and interlock.
▶ when the generator is a diesel engine driven set or rotary UPS as well as the forgoing it is always wise to disconnect the engine starting battery and lock off the battery leads and place caution signs as an extra safeguard.

▼ **Figure 4.11** Lead lock-out bag – can also be used to ensure security in isolation for plug-in devices

514.15.1 Where an installation contains a generator arranged to provide an additional source of supply in parallel with another source, it is necessary to post the warning notice shown in Figure 4.12 in the following locations within the installation to raise awareness of the need to isolate more than one source of supply:

▶ at the origin of the installation
▶ at the meter position, if remote from the origin
▶ at the distribution board or consumer unit to which the generator is connected
▶ at ALL points of isolation provided for ALL sources of supply.

▼ **Figure 4.12** Example of multiple supplies warning label (Regulation 514.15.1)

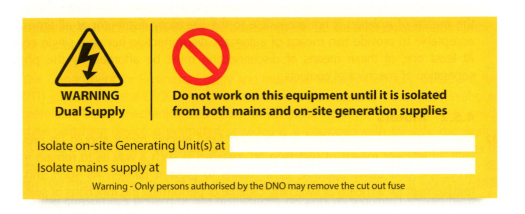

WARNING
Dual Supply

Do not work on this equipment until it is isolated from both mains and on-site generation supplies

Isolate on-site Generating Unit(s) at

Isolate mains supply at

Warning - Only persons authorised by the DNO may remove the cut out fuse

It is becoming increasingly common for Small-Scale Embedded Generators (SSEG) to be installed within even small installations such as domestic premises.

Where this is the case, it is necessary to provide a means of:

551.7.4 ▶ automatic switching that will disconnect the generator from the distribution network operator's (DNO's) distribution system in the event of the public supply failing or exceeding the acceptable variation of voltage and frequency that are given in the Energy Networks Association (ENA) document EREC G83/2-1 or G98.

551.7.6 ▶ isolating the generator from the DNO distribution system. For a generating set with an output exceeding 16 A, the accessibility of this means of isolation is to comply with the wiring regulations and the distribution system operator (DNO) requirements. For a generating set with an output not exceeding 16 A, the accessibility of this means of isolation is to comply with BS EN 50438:2013 (this standard is currently being revised).

▼ **Figure 4.13** Isolation arrangements for small-scale generator

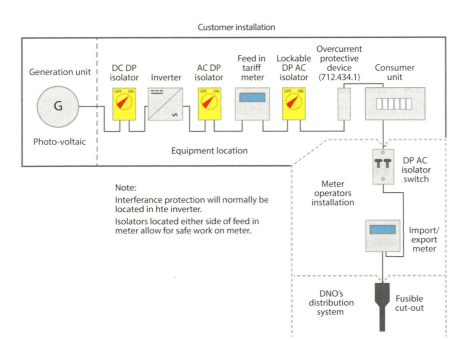

Note:
Interferance protection will normally be located in hte inverter.
Isolators located either side of feed in meter allow for safe work on meter.

Previously BS 5839 had required a DP switch on the mains supply local to the fire alarm control panel, but this is no longer required in BS 5839:2017 and fire alarm installation personnel must be trained in safe isolation procedures as are all other electrical operatives (see Appendix B for further guidance).

Switching off for mechanical maintenance

5

Before considering the requirements for mechanical maintenance in BS 7671 it should be noted that where electrical equipment falls within the scope of BS EN 60204 *Safety of Machinery. Electrical equipment of machines*, the requirements for mechanical maintenance contained therein will apply.

5.1 General

BS 7671 only describes requirements for 'electrical' switching-off. It should be remembered, however, that many items of industrial equipment may involve electrically powered or actuated parts. A risk may also arise from items of equipment containing electromagnetic operation or electrical heating elements. As such it may be necessary to provide other, non-electrical, measures to prevent danger. By way of example, braking bars or similar movement-limiting devices might need to be installed to prevent the possibility of parts moving or falling. Where this is the case, any such precautions required should be clearly stated in the operational manual for that part of the installation.

537.3.2.2 Wherever mechanical maintenance work might involve a risk of physical injury, a means of switching off should be provided to facilitate such maintenance work being carried out safely.

462.3
464.2 Similar to the requirements for isolation described earlier, except where the means of switching off remains under the control of the persons dependent upon it remaining effective, it is necessary to provide a means of preventing electrically powered or operated equipment from becoming reactivated unintentionally whilst mechanical maintenance work is taking place.

5.2 Devices for switching off for mechanical maintenance

537.3.2.2 Any device provided for the purpose of switching off for mechanical maintenance should preferably be placed in the main supply circuit so as to work as directly as possible. Any switch so provided should be capable of being operated on-load at the full load current and hence be suitable for use by ordinary persons. It need not interrupt the neutral conductor for some supply arrangements; see Sections 4.2 and 4.3 of this Guidance Note.

If switching off for mechanical maintenance is achieved by interrupting the control circuit of a drive, either mechanical restraints should also be employed or the requirements of the relevant British Standard specification for the control device employed.

537.3.2.3 Whatever device is selected to provide switching off for mechanical maintenance it should require manual operation. This would include a control switch acting on a contactor.

Again, as was the case for isolation, where the switching contacts cannot be seen to be in the OPEN position, a reliable means of indicating the OPEN position should be provided. This is achieved typically by 'O' and 'I' symbols indicating the OPEN (off) and CLOSED (on) positions respectively.

537.3.2.4 The means of switching-off must be so designed or installed that it cannot be switched back on inadvertently, unintentionally, or prematurely. Its purpose must be clear either by position or by labelling and it should be placed so as to be easily used as and when required. The use of locks is often required here.

The use of a plug and socket-outlet arrangement is now not specifically permitted for use as the means of providing switching off for mechanical maintenance but it is not specifically precluded either, and it is often reassuring for a person who is not electrically competent to see that the means of supply is so obviously disconnected. There are many machines such as lawnmowers, vacuum cleaners etc. where adjustments or cleaning regularly need to be carried out, and as long as persons are trained in the correct use of the equipment and unplug the equipment and maintain control of the plug such switching is likely to be acceptable.

Emergency switching and emergency stopping — 6

Before considering the requirements for emergency switching in BS 7671 it should be noted that where electrical equipment falls within the scope of BS EN 60204 *Safety of Machinery. Electrical equipment of machines*, the requirements for emergency switching contained therein will apply.

6.1 General

465.4 It should be noted, as was stated in Section 3.4 of this Guidance Note, that emergency switching may cause a supply to be disconnected or energized. Operation of an emergency switch might, for example, activate extraction plant or instigate the release of extinguishing agents. Operation of an emergency stop device might cause electromechanical braking to operate as a result of releasing a motor which was being held-off when supplied.

465.1 Emergency switching arrangements should be provided as and where required within an installation to control the electrical supply to circuits or equipment to remove a danger that has occurred unexpectedly.

465.2 In an installation supplied from a TN system, emergency switching devices need not
465.3 interrupt the neutral conductor and it is preferred that they act as directly as possible on the relevant supply conductors. That is not to say, however, that emergency stopping devices cannot act upon the control circuit of, for example, a contactor.

465.4 However the means of emergency switching is arranged, its operation must not introduce other hazards or interfere with any operation initiated to remove danger. This factor needs to be borne in mind when considering the release of extinguishing agents into a switchroom for example.

6.2 Devices for emergency switching

537.3.3.2 Any device selected to act as an emergency switch should be capable of breaking
537.3.3.5 the full load current likely to be present at the point of the installation where it is
537.3.3.6 installed, including the current that would flow under locked rotor conditions of a motor. Emergency stop buttons are commonly arranged to interrupt a control circuit which, when interrupted, causes a contactor to operate. Where a contactor or circuit-breaker is operated remotely, it should open when its operating coil is de-energized, although other methods providing the same degree of reliability are not excluded.

537.3.3.3 Such an arrangement is permitted by Regulation 537.3.3.3, which states that the means of emergency switching may be:

- ▶ a single device acting directly on the supply, or
- ▶ a combination of equipment activated by a single action.

537.3.3.5
537.3.3.6
BS EN ISO 13850 It is preferred that emergency switching devices be of a hand-operated type, their operating button or handle being clearly identifiable by colour. If colour is used it should be red against a contrasting background. Clause 4.4.5 of BS EN ISO 13850 *Safety of Machinery. Emergency stop. Principles for design* states that, as far as it is practicable, the background shall be coloured YELLOW, see Figure 6.1.

▼ **Figure 6.1** Examples of emergency stop buttons [photograph courtesy of Eaton Moeller]

537.3.3.6 Emergency switching/stopping devices should be installed in locations where danger might occur and, where necessary, in any number of remote positions such that danger can be removed rapidly. This requirement would apply, for example, to a typical production line or an escalator.

537.3.3.7 Except for situations where the means of emergency switching and re-energizing are both under the control of the same person, the means of emergency switching should be capable of latching or otherwise being restrained in the OFF or STOP position.

537.3.3.5 As they are provided as a means of rapidly removing a source of danger, any installed emergency switching/stopping device should be clearly identifiable so that it can be easily located in an emergency and should remain readily accessible whenever the installation or part thereof controlled by the device is in use.

Table 537.4 Table 3.1 of this Guidance Note states that a plug and socket-outlet arrangement must not be selected for use as a means of providing emergency switching.

Where there is a plug and socket-outlet in addition to the means of emergency switching, the regulations do not preclude the use of unplugging from that socket-outlet as a means of disconnecting the supply in an emergency if it is easily accessible. If that plug and socket-outlet could be used in this way, the designer would need to consider the relative risks involved; for example, the act of unplugging could cause sparks or it may interfere with a designed emergency stopping operation such as DC braking or other automated shutdown procedures.

Functional switching 7

7.1 General

463.1.1
463.1.2
463.3
537.3.3.3
Functional switching arrangements should be provided wherever it is necessary for a circuit or part thereof to be separately controlled. The type and number of switches so selected should be appropriate for the purpose. However, functional switching devices need not act upon all live conductors of a circuit. The use of the term 'act upon' here is deliberate, as functional (control) switching devices are not confined solely to two-state (i.e. on/off) switches. Dimmer switches and programmable speed controllers are examples of variable control devices.

It is acceptable for a single functional switching device to be used to control more than one item of equipment.

463.1.4
Where a functional switching device is provided to allow the changeover of supply between alternative sources, the switching device should interrupt all live conductors and should be arranged such that the sources of supply cannot operate in parallel unless they are designed to be capable of so doing.

461.2
543.3.3.101
The requirements of Regulation 461.2 and of Regulation 543.3.3.101, that combined protective earth and neutral (PEN) conductors and protective conductors respectively should not be switched, apply to functional switching.

7.2 Functional switching devices

537.3.2.1
Functional switching devices must be suitable for the most onerous duty that they will need to perform within the particular design constraints. Design constraints would include the suitability to switch the type of load (fluorescent / ELV halogen lamp, motor, etc.) and meet the required number of operations, for example, frequent use. For examples of typical applications see Table 3.1.

A device embodying more than one function should comply with all the requirements appropriate to each separate function for protection, isolation, switching, control and monitoring.

Circuit-breakers to BS EN 60898-1 and BS EN 60947-2 and RCDs to BS EN 61008, BS EN 61009 and BS EN 60947-2 are capable of functional switching as indicated in Table 3.1. However, circuit-breakers and RCDs are primarily circuit protective devices and, as such, they are not intended for frequent load switching.

For a more frequent duty, the number of operations and load characteristics according to the manufacturer's instructions should be taken into account or an alternative device from those listed as suitable for functional switching in Table 537.4 should be employed.

537.2.2 Although semiconductors are not permitted to be used as a means of isolation, they may be used for functional switching purposes.

Part L of the Building Regulations in England and Wales covers the conservation of fuel and power and as lighting uses electrical energy its consumption will need to be accounted for in building energy design and management systems and suitable controls provided. Also modifications to lighting systems are a controlled service and need to comply with the requirements of the Building Regulations.

In Scotland the legislation is The Building (Scotland) Regulations 2004 as amended. The legislation applies generally to buildings and work carried out in buildings in Scotland with some exemptions. Compliance with the building regulations is achieved by fulfilling the requirements of the mandatory building standards and guidance on compliance with these is provided in two technical handbooks produced by the Scottish Government Building Standards Division. There is a handbook for domestic premises and another for non-domestic premises.

7.3 Control (auxiliary) circuits

557.3.201 Control circuits should be designed, arranged and protected in such a way as to prevent the equipment being controlled from operating inadvertently in the event of a fault. It is good practice for control systems to be arranged as simply as possible while being fail-safe.

7.4 Motor control

463.3 Before considering the requirements for motor control in BS 7671 it should be noted that where electrical equipment falls within the scope of BS EN 60204 *Safety of Machinery. Electrical equipment of machines*, the requirements for mechanical maintenance contained therein will apply.

552.1.2
552.1.3 Except where it can be shown that a failure to restart would result in greater danger, control measures should be put in place to prevent a motor from automatically restarting after stopping as a result of voltage drop or temporary supply loss.

463.3.2
463.3.3 If, as is sometimes the case, a motor is subject to reverse-current braking, and where such reversal might result in danger, measures should be taken to prevent the reversal of direction of rotation after the driven parts come to a standstill at the end of the braking period. Further, where safety is dependent upon the motor operating in the correct direction, means should be provided to prevent reverse operation.

Firefighters' switches 8

8.1 General

537.4 Requirements relating to firefighters' switches are contained in Regulation group 537.4, comprising of six regulations.

The Regulatory Reform (Fire Safety) Order 2005 (henceforth referred to as the Order) has a direct influence on many other pieces of primary and secondary legislation, requiring modifications and, in some cases, partial or full revocation of the requirements therein. It replaced fire certification under the Fire Precautions Act 1971 with a general duty to ensure, so far as is reasonably practicable, the safety of employees and a general duty, in relation to non-employees, to take such fire precautions as may reasonably be required to ensure that premises are safe.

The Order contained a number of statutory requirements relating to firefighters' switches for luminous tube signs, etc., which are self-contained apparatus consisting of luminous tube signs designed to work at a voltage normally exceeding the prescribed low voltage level specified in BS 7671.

Article 37(4) of the Order states that a firefighters' switch must be so placed and coloured or marked as to satisfy such reasonable requirements that a fire and rescue authority may impose so that it is readily recognisable by and accessible to firefighters.

Although the detailed requirements of BS 7671 are of themselves non-statutory, article 37(5) of the Order makes it clear that if a firefighters' switch complies in position, colour and marking with the current edition of the IET Wiring Regulations, fire and rescue authorities should not impose any further requirements on such matters. It can be seen therefore that meeting the requirements of BS 7671 for the selection and installation of firefighters' switches will in itself meet the relevant statutory requirements of the relevant fire safety legislation.

In Northern Ireland, The Fire Safety Regulations (Northern Ireland) 2010 apply and in Scotland, The Fire Safety (Scotland) Regulations 2006 apply.

The requirements of BS 7671:2018 relating to firefighters' switches are summarised on the next page.

537.4.1
537.4.2 A firefighters' switch is to comply with BS EN 60669-2-6 or BS EN 60947-3 and be provided on the low voltage side of a circuit that supplies

▶ any exterior electrical installation, or
▶ an interior discharge lighting installation (not including a portable luminaire or sign of rating not exceeding 100 W)

that operates at a voltage in excess of low voltage.

It should be noted at this point that in relation to the provision of firefighters' switches, installations in covered markets, arcades or shopping malls are considered to be exterior installations, while a temporary installation in a permanent building intended for hosting exhibitions is not. It is also to be noted that in certain premises subject to licensing conditions, such as petrol station forecourts, the licencing authority may require the installation of a firefighter's switch.

537.4.2.1 Wherever practically possible, every exterior installation in one premises covered by the requirements of Regulation 537.4.2.1, should be controlled by a single firefighters' switch (Figure 8.1) to simplify the process of making dead all circuits therein operating at a voltage exceeding low voltage. Further, every internal installation subject to the requirements of Regulation 537.4.2.1 in each individual premises should be controlled by a single firefighters' switch, this switch being independent of a firefighters' switch for any exterior installation of the same premises.

8.2 Location and identification

537.4.2.2 Any firefighters' switch provided should, in the case of an exterior installation, be outside the building. Alternatively, where the switch and equipment being controlled are not adjacent to each other, notices should be placed adjacent to both the switch, indicating what it controls, and the equipment controlled, indicating the location of the appropriate switch.

For an interior installation, the switch should be placed in the main entrance of the building or other location agreed to by the local fire authority.

▼ **Figure 8.1** Surface mounted firefighters' switch (BS EN 60669-2-6 also allows the colour to be black)

In all cases, the switch should be in a conspicuous position that remains accessible to firefighters. The switch should be mounted at a height of not more than 2.75 m from the ground or the standing beneath the switch.

Where more than one switch is provided for any one building, notices are required clearly describing the installation or part thereof which each switch controls.

537.4.3 Any firefighters' switch provided should be easily visible and so placed to facilitate operation by a firefighter and marked to indicate the installation or part of the installation which it controls.

537.4.4 Each firefighters switch is to be:

▶ be coloured red
▶ be accompanied by a durable notice with the wording as shown in Figure 8.2 (it is not required for the sign to be red but it may be helpful to have the sign the same colour as the firefighters switch)
▶ have its ON and OFF positions clearly marked such as to be legible to a person standing in a position to operate the switch
▶ have its OFF position uppermost
▶ be so constructed such that it cannot be inadvertently switched to the ON position.

▼ **Figure 8.2** Example of firefighter's switch label called for by Regulation 537.4.4

9.1 General

110.1
Part 7
Part 7 of BS 7671 contains requirements for twenty types of special installation or location. Of these twenty sections, seven contain specific requirements falling within the remit of this Guidance Note. These are summarised below.

9.2 Construction sites

Sect 704
704.537.2
Section 704 gives requirements for construction and demolition site installations. Regulation 704.537.2 requires that each Assembly for Construction Sites (ACS) incorporates suitable means of isolating the incoming supply to the temporary installation.

Other standards pertinent to construction sites are:

▶ BS 4363:1998+A1:2013 *Specification for distribution assemblies for reduced low voltage electricity supplies for construction and building sites*
▶ BS 7375:2010 *Distribution of electricity on construction and building sites – Code of practice*
▶ BS EN 60439-4:2013 *Low voltage switchgear and controlgear assemblies. Particular requirements for assemblies for construction sites (ACS).*

Clause 7.1.2 of BS 7375 requires that only multi-pole isolators are used, breaking all line conductors simultaneously (including the neutral conductor), due to the likelihood of a TT system being used. This means of isolation should be capable of being secured in the OFF position by means of a key or special tool (unless it is of a plug and socket type).

Clause 101.2 of BS EN 60439-4 requires any incoming unit to include a means of isolation that can be secured in the OPEN position.

9.3 Agricultural and horticultural premises

Sect 705
705.53
It is a requirement for any electrical heating appliance selected for use within an agricultural or horticultural installation to incorporate a visual indication of whether it is in the ON or OFF position.

705.537.2
It is a requirement for the electrical installation of each building or part of a building to be provided with a single means of isolation.

Some items of electrical equipment in agricultural and horticultural premises remain subject to seasonal patterns of usage connected for example to harvesting times. Where this is the case, a means of isolation acting on all live conductors, including the neutral conductor, should be provided.

705.537.2 Isolation and switching devices installed within agricultural or horticultural premises should be so placed as to be out of the reach of livestock to prevent damage or unintended operation and such that they will not be made inaccessible when required. Furthermore, a clear indication should be given of the purpose of these isolating devices where this is not immediately apparent.

9.4 Marinas

Sect 709 Section 709 contains requirements applicable to electrical installations in marinas.

709.537.2.1.1
709.411.4 Note
709.537.2.1.1 It is commonplace in marina installations for distribution cabinets to be provided containing the equipment necessary for the connection of a number of vessels. Regulation 709.537.2.1.1 requires that each distribution cabinet should be provided with a means of isolation. The Electricity Safety, Quality and Continuity Regulations effectively prohibit the use of a PME earthing facility for electrical supplies to boats. Isolators installed in marina installations must disconnect all live conductors including the neutral conductor. It is permitted for one isolator to serve up to four socket-outlets (see Figures 9.1 and 9.2).

▼ **Figure 9.1** Example of the means of connection for use with vessels in marinas (single-phase)

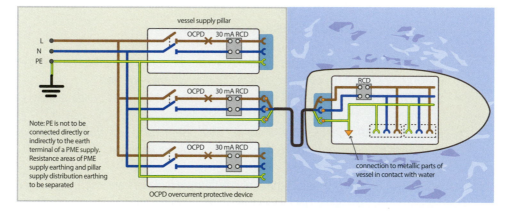

▼ **Figure 9.2** Example of the means of connection for use with vessels in marinas (three-phase)

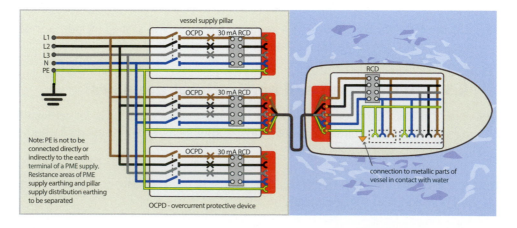

9.5   Exhibitions, shows and stands

Sect 711   Section 711 relates to installations for exhibitions, shows and stands.

711.537.2.3   Regulation 711.537.2.3 requires that each separate temporary structure, vehicle, stand or unit which is intended to be occupied by one user and each distribution circuit being used to provide a supply to outdoor installations of this type should be provided with its own means of isolation. The means of isolation should be readily accessible and clearly identifiable.

711.55.4.1   Where electric motors are provided and a hazardous situation might develop, a means of isolation acting on all poles of its supply should be provided. Either the isolator or its means of control, say a lock-stop device or similar, should be adjacent to the motor.

711.559.4.4.3   Where the installation of an exhibition, show or stand contains discharge lighting operating at a voltage in excess of 230/400 V AC this should be supplied from a separate circuit which is controlled by an emergency switch which itself should be easily visible, accessible and clearly identified.

711.410.3   Automatic disconnection of the supply to a temporary electrical installation for a booth or stand etc. is to be provided at the origin by the use of a 300 mA (or less) RCD. Selectivity by time delay is provided where required with final circuits.

9.6   Solar photovoltaic (PV) power supply systems

Sect 712   Section 712 contains requirements for this type of installation.

551.7.6   For a solar PV installation designed to operate in parallel with the public supply, reference should be made to Regulation 551.7.6 and the guidance on this regulation given in Section 4.5.1 of this Guidance Note.

712.410.3   There are particular problems associated with solar PV power supply systems in that it is difficult to remove all sources of light acting upon the solar PV arrays (solar panels). This is recognised in Regulation 712.410.3, which states that PV equipment on the DC side shall be considered to be energized, even when the system is disconnected from the AC side.

712.537.2.1.1   As a means of facilitating maintenance of the PV convertor (invertor), isolation devices
Fig 712.1   must be provided to allow the PV convertor to be disconnected from both the DC side
712.537.2.2.5   and the AC side. This arrangement is illustrated in Figure 712.1 of BS 7671. It should be noted at this point that Regulation 712.537.2.2.5 also requires the installation of a switch-disconnector to be installed on the DC side of the convertor.

712.537.2.2.1   When considering the selection and erection of devices for isolation and switching for use between the solar PV supply system and the public supply network, the public supply should be considered to be the source (as defined) and the PV installation should be considered to be the load.

712.537.2.2.5.1   As a result of the difficulties involved in confirming that solar PV systems are fully isolated, as mentioned above, it is a requirement to post warning signs on all junction boxes employed within the solar PV system, pointing out that parts within the boxes might still be live after isolation from the PV convertor (Figure 9.3).

▼ Figure 9.3 Example of warning label required for junction boxes in solar PV systems

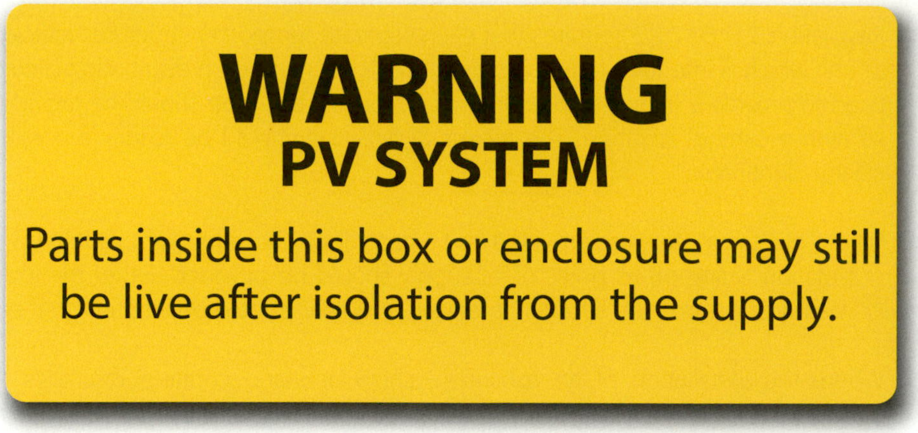

9.7   Caravans and motor caravans

Sect 721   Section 721 contains particular requirements to be applied to electrical installations in caravans and motor caravans.

721.537.2.1.1   Each individual caravan or motor caravan should be provided with a main isolator arranged to disconnect all live conductors. It should be placed to allow for ready operation. If the caravan consists of only one final circuit, the overcurrent protective device for the circuit may be used as the isolator. It is usual for this isolator to be an RCD that provides additional protection.

721.537.2.1.1.1   Regulation 721.537.2.1.1.1 requires the installation of an advisory notice close to the
Fig 721   main isolating switch. The wording required for this notice is given in Figure 721 of BS 7671. This includes an explanation of how to disconnect the caravan making use of the main isolating switch.

9.8   Structures, amusement devices and booths at fairgrounds, amusement parks and circuses

Sect 740   Section 740 contains requirements applicable to electrical installations of structures, amusement devices and booths at fairgrounds and the like.

740.537.1   Such installations are likely to comprise a number of separate structures, booths, rides
740.537.2.1.1   or similar. Regulation 740.537.1 requires each booth, structure, ride or similar to be provided with its own means of isolation which should remain readily accessible at all times. A separate means of isolation is also required for each distribution circuit intended to supply outdoor installations.

740.537.2.2   All switching devices selected to act as the means of isolation in such locations should disconnect all live conductors including the neutral conductor.

740.55.3.2   Any luminous tubes or signs which operate at a voltage exceeding 230/400 V AC should be connected to a circuit which is controlled by an emergency switching arrangement (see Chapter 8).

740.410.3   Automatic disconnection of the supply to a temporary electrical installation for a booth or stand etc. is to be provided at the origin by the use of a 300 mA (or less) RCD. Selectivity by time delay is provided where required with final circuits.

Appendix A

Safety service and product standards of relevance to this Guidance Note

BS or BS EN number	Title	References in this Guidance Note
BS 7671:2018	Requirements for electrical installations. IET Wiring Regulations. Eighteenth Edition	Numerous
BS 88 series	Cartridge fuses for voltages up to and including 1000 V AC and 1500 V DC	Table 3.1
BS 1362:1973	Specification for general purpose fuse links for domestic and similar purposes (primarily for use in plugs)	Table 3.1
BS 1363 series	13 A plugs, socket-outlets, adaptors and connection units	3.1
BS 1363-1:2016+A1:2018	13 A plugs, socket-outlets, adaptors and connection units. Specification for rewirable and non-rewirable 13 A fused plugs	Table 3.1
BS 1363-2:2016+A1:2018	13 A plugs, socket-outlets, adaptors and connection units. Specification for 13 A switched and unswitched socket-outlets	Table 3.1
BS 1363-4:2016+A1:2018	13 A plugs, socket-outlets, adaptors and connection units. Specification for 13 A fused connection units switched and unswitched	Table 3.1
BS 4177:1992	Specification for cooker control units	Table 3.1
BS 4363:1998+A1:2013	Specification for distribution assemblies for reduced low voltage electricity supplies for construction and building sites	9.2
Replaced by BS EN 82079-1	Preparation of instructions for use. Structuring, content and presentation. General principles and detailed requirements	Introduction
BS 4940 series	Technical information on constructional products and services	Introduction
BS 5266-1:2016	Emergency lighting. Code of practice for the emergency lighting of premises	4.3
BS EN 60669-2-1:2004+A12:2010	Specification for electronic variable control switches (dimmer switches) for tungsten filament lighting	
BS 5733:2010+A1:2014	General requirements for electrical accessories. Specification	Table 3.1

BS or BS EN number	Title	References in this Guidance Note
BS 5839-1:2002+A2:2008	Fire detection and fire alarm systems for buildings. Code of practice for system design, installation, commissioning and maintenance	4.3
BS 5839-6:2013	Fire detection and fire alarm systems for buildings. Code of practice for system design, installation, commissioning and maintenance	4.3
BS 6972:1988	Specification for general requirements for luminaire supporting couplers for domestic, light industrial and commercial use	Table 3.1
BS 7375:2010	Distribution of electricity on construction and demolition sites. Code of practice	9.2
BS 7671:2018	Requirements for electrical installations. IET Wiring Regulations. Eighteenth Edition	Numerous
BS EN 50428:2005+A2:2009	Switches for household and similar fixed electrical installations. Collateral standard. Switches and related accessories for use in home and building electronic systems (HBES)	3.1; Table 3.1
BS EN 50438:2013	Requirements for the connection of micro-generators in parallel with public low-voltage distribution networks	4.5.1
BS EN 60073:2002	Basic and safety principles for man-machine interface, marking and identification. Coding principles for indicators and actuators	3.1
BS EN 60204 series	Safety of machinery. Electrical equipment of machines	1.2.4; 5; 6; 7.4
BS EN 60204-1:2006+A1:2009	Safety of machinery. Electrical equipment of machines. General requirements	3.5
BS EN 60309 series	Plugs, socket-outlets and couplers for industrial purposes	Table 3.1
BS EN 60309-1:1999+A2:2012	Plugs, socket-outlets and couplers for industrial purposes. General requirements	3.1
BS EN 61439-4:2013	Low-voltage switchgear and controlgear assemblies. Particular requirements for assemblies for construction sites (ACS)	9.2
BS EN 60669-1:1999+A2:2008	Switches for household and similar fixed-electrical installations. General requirements	3.1; Table 3.1
BS EN 60669-2-series	Switches for household and similar fixed-electrical installations. Particular requirements	3.1
BS EN 60669-2-1:2004+A12:2010	Switches for household and similar fixed-electrical installations. Particular requirements. Electronic switches	3.1; Table 3.1
BS EN 60669-2-2:2006	Switches for household and similar fixed-electrical installations. Particular requirements. Electromagnetic remote-control switches (RCS)	3.1; Table 3.1

BS or BS EN number	Title	References in this Guidance Note
BS EN 60669-2-3:2006	Switches for household and similar fixed-electrical installations. Particular requirements. Time-delay switches (TDS)	3.1; Table 3.1
BS EN 60669-2-4:2005	Particular requirements. Isolating switches	3.1; Table 3.1
BS EN 60898 series	Circuit breakers for overcurrent protection for household and similar installations	Table 3.1
BS EN 60898-1:2003+A13:2012	Circuit breakers for overcurrent protection for household and similar installations. Circuit-breakers for AC operation	7.2
BS EN 60947 series	Low-voltage switchgear and controlgear	3.1; Table 3.2
BS EN 60947-1:2007+A1:2011	Low-voltage switchgear and controlgear. General rules	3.8
BS EN 60947-2:2006+A2:2013	Low-voltage switchgear and controlgear Circuit-breakers	Table 3.1; Table 3.27.2
BS EN 60947-3:2009+A1:2012	Low-voltage switchgear and controlgear. Switches, disconnectors, switch-disconnectors and fuse-combination units	Table 3.1; Table 3.2
BS EN 60947-4-1:2010+A2:2014	Low-voltage switchgear and controlgear. Contactors and motor-starters. Electromechanical contactors and motor-starters	Table 3.1; Table 3.2
BS EN 60947-4-2:2012	Low-voltage switchgear and controlgear. Contactors and motor-starters. AC semiconductor motor controllers and starters	Table 3.2
BS EN 60947-4-3:2014	Low-voltage switchgear and controlgear. Switches, disconnectors, switch-disconnectors and fuse-combination units	Table 3.2
BS EN 60947-5-1:2017	Low-voltage switchgear and controlgear. Control circuit devices and switching elements. Electromechanical control circuit devices	Table 3.1; Table 3.2
BS EN 60947-5-2:2007+A1:2012	Specification for low-voltage switchgear and controlgear. Control circuit devices and switching elements. Proximity switches	Table 3.2
BS EN 60947-5-5:1998+A2:2017	Low-voltage switchgear and controlgear. Control circuit devices and switching elements. Electrical emergency stop devices with mechanical latching function	3.5
BS EN 60947-6-1:2005+A1:2014	Low-voltage switchgear and controlgear. Multiple function equipment. Transfer switching equipment	Table 3.1; Table 3.2
BS EN 60947-6-2:2003	Low-voltage switchgear and controlgear. Multiple function equipment. Control and protective switching devices (or equipment) (CPS)	Table 3.1; Table 3.2

BS or BS EN number	Title	References in this Guidance Note
BS EN 61008 series	Residual current operated circuit-breakers without integral overcurrent protection for household and similar uses(RCCBs)	7.2
BS EN 61008-1:2012+A12:2017	Residual current operated circuit-breakers without integral overcurrent protection for household and similar uses (RCCBs). General rules	Table 3.1
BS EN 61009 series	Residual current operated circuit-breakers with integral overcurrent protection for household and similar uses (RCBOs)	7.2
BS EN 61009-1:2012+A12:2016	Residual current operated circuit-breakers with integral overcurrent protection for household and similar uses (RCBOs). General rules	Table 3.1
BS EN 61095:2009	Electromechanical contactors for household and similar purposes	Table 3.1; Table 3.2
BS EN 61995-1:2008	Devices for the connection of luminaires for household and similar purposes. General requirements	Table 3.1
BS EN ISO 12100:2010	Safety of machinery. General principles for design. Risk assessment and risk reduction	1.2.4
BS EN ISO 13850:2008	Safety of machinery. Emergency stop. Principles for design	1.2.4; 3.5; 6.2; Figure 6.1
IEC 60050	International electrotechnical vocabulary. Switchgear, controlgear and fuses. (note the quoted definition IEV 441-17-20 was taken from IEC 60050)	3.83.8
IEC 60884 series	Plugs and socket-outlets for household and similar purposes	Table 3.1
IEC 60906 series	IEC system of plugs and socket-outlets for household and similar purposes	Table 3.1

Appendix
Safe isolation procedures

B

This appendix is not intended to be an exhaustive treatment of the subject of safe isolation, but rather a reminder of the minimum steps which should be taken to confirm that an installation, circuit or item of equipment has been isolated and made dead. References are given at the end of this appendix for further sources of information regarding safe isolation.

Minimum stages to confirm safe isolation:

▶ Before project start date plan work and agree method statement with client or duty holder
▶ On project start date confirm with client or duty holder that the proposed work can be carried out
▶ Locate all sources of supply to the installation or circuit such as standby generators, PV systems, UPSs, secondary sources of supply, CHP systems etc. and isolate those supplies if necessary. Consider whether the isolated circuit could become live at any time due to secondary sources of supply or through switching on another circuit with a 'borrowed neutral' from the isolated circuit. If there is any doubt, it may be necessary to isolate the whole installation at the incoming supply position. Where power factor correction units or other capacitors are installed these will need to be discharged or a suitable time elapsed before testing commences
▶ Locate/positively identify correct circuit and isolation point or device
▶ Select an approved voltage tester and check for damage or deterioration. Preferably use an approved voltage tester that does not require a potential difference between the probes to give a warning. Some testers require a potential difference between the 2 probes to give an indication of a live part, a terminal may still be live but without a potential difference present there will be no indication of the live part
▶ Confirm that voltage indicating device is functioning correctly against any know source
▶ Switch off installation/circuit to be isolated
▶ Verify with voltage indicating device that no voltage is present at point of isolation by checking between earth and all live terminals, between neutral and live terminals and between line terminals
▶ Re-confirm that voltage indicating device functions correctly on known supply or proving unit
▶ Lock off or otherwise secure device used to isolate installation/circuit and label to identify who is to responsible for isolation
▶ Provide effective control of any area where work is being undertaken, e.g. personnel barriers around area of isolation and area of proposed work ensuring that fire exits and safety services provisions are accessible
▶ Post warning and emergency contact sign(s)

▶ Prove dead at point of proposed work before work begins by checking between earth and all live terminals, between neutral and live terminals and between line terminals

▶ Issue 'Permit to Work' if applicable.

In order to identify appropriate points or devices to carry out the isolation, it is important to employ all relevant sources of information. These would include discussions with the persons responsible for the electrical installation and perusal of building operation manuals, installation schematic drawings and circuit schedules.

It is highly recommended that a voltage indicator or test lamp, which has been expressly designed for proving the absence of supply is employed when carrying out safe isolation. These have a number of design features to protect the user including integral current-limiting measures; finger guards; appropriately insulated leads; and probes arranged such that a minimum (2 to 4 mm) of uninsulated tip is exposed. The use of multi-range instruments is not recommended. They may lack a number of the safety features of a voltage indicator mentioned above, and incorrect ranges may be selected, both of which may place a user in danger.

The voltage indicator/test lamp should be checked for signs of damage or deterioration prior to each use. Any voltage indicator/test lamp with significant damage resulting, for example, in exposed conductive parts which will be live during use should be removed from service immediately and if not repairable be destroyed to prevent its unintended or inadvertent use.

It is essential that the user confirms the functionality of the voltage indicator/test lamp before EACH use. This may be achieved by proving on a known supply or proprietary proving unit, even where the voltage indicator has an integral 'self-test' facility.

After switching off the installation, circuit or item of equipment to be isolated, the 'proved' voltage indicator/test lamp should be used to confirm the absence of any voltage by testing sequentially between all conductors – phase to phase (in three-phase systems), each phase to neutral, each phase to earth and neutral to earth. The correct operation of the voltage indicator/test lamp should be reconfirmed using a known supply or proving unit between each test. Beware of 'borrowed neutrals' from other circuits as that can make a circuit live even though their supply is isolated (see Figure B.1).

▼ **Figure B.1**   Borrowed neutral

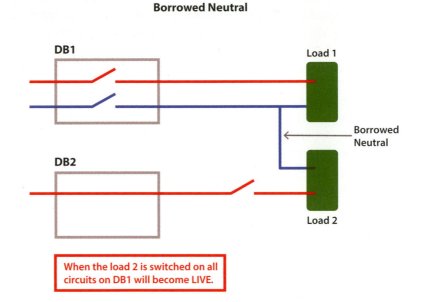

Borrowed Neutral

DB1

Load 1

Borrowed
Neutral

DB2

Load 2

**When the load 2 is switched on all
circuits on DB1 will become LIVE.**

Having confirmed that the correct installation, circuit or item of equipment has been made dead, the means of isolation should be made secure. Unless the means of isolation remains under the control of the person(s) dependent upon it, this will involve the use of some means of locking off. A wide range of lock-out devices are available allowing virtually any type of device (circuit-breaker, switch, plug, etc.) to be secured against inadvertent or unauthorised use. It is never acceptable to merely place a strip of tape over a device which has been switched off.

Any warning sign posted should convey in simple terms that the installation, circuit or item of equipment has been deliberately disconnected from the supply (see Figure B.2). It should inform the reader that care must be taken to ensure the safety of persons who are reliant upon the continued effectiveness of the isolation and should state that the installation, circuit or equipment should not be re-energized until it has been confirmed that it is safe to do so.

The legal requirements relating to safe isolation procedures can be found in Regulation 12 'Means for cutting off the supply and for isolation' and Regulation 13 'Precautions for work on equipment made dead' of The Electricity at Work Regulations (EWR) 1989.

Employers, employees and the self-employed working on or near electrical installations, or responsible for such work, should be familiar with the statutory requirements imposed on them by the EWR 1989.

The Health and Safety Executive (HSE) has produced *The Electricity at Work Regulations 1989. Guidance on Regulations HSR25* to clarify the requirements of the EWR 1989. It is essential reading for all persons upon whom the EWR 1989 impose duties.

Another HSE publication *Electrical test equipment for use by electricians* (GS 38) gives guidance on the selection and use of voltage indicating devices to be used when carrying out safe isolation procedures. This too is essential reading.

The HSE publication *Electricity at work. Safe working practices* (HSG85) gives a broader overview of all aspects of working on or near electrical installations.

Before any work can be carried out it must be properly planned and the HSE site (hse. gov.uk/construction/safetytopics/admin.htm) provides extensive guidance on this. The law on construction health and safety requires action to protect those at work on site and members of the public who may be affected. The key safety and health topics which require attention are covered in these webpages:

▶ HSE (www.hse.gov.uk) contains various publications including a considerable quantity of free downloadable publications including HSR25 and HSG85. HSE publications INDG372 and HSG230 provide guidance on the long term management of LV electrical switchgear
▶ Electrical Safety First (www.electricalsafetyfirst.org.uk/electrical-professionals/ best-practice-guides/) *Best Practice Guide No. 2 (Issue 3) Guidance on the management of electrical safety and safe isolation procedures for low voltage installations.*

▼ **Figure B.2** Examples of safe isolation warning sign

DO NOT ATTEMPT TO REMOVE

THIS SOURCE OF ELECTRICAL SUPPLY HAS BEEN DELIBERATELY ISOLATED AND SECURED.

IET
Electrical excellence
www.theiet.org/wiringregs

MINIMUM STAGES FOR SAFE ISOLATION

- Locate/positively identify correct isolation point or device
- Check condition of voltage indicating device
- Confirm that voltage indicating device is functioning correctly
- Switch off installation/circuit to be isolated
- Verify with voltage indicating device that no voltage is present
- Re-confirm that voltage indicating device functions correctly on known supply/proving unit
- Lock-off or otherwise secure device used to isolate installation /circuit
- Post warning sign(s)

Along with this label there must also be some indication of the person in charge of the work and a contact means in case of emergencies. Also it is helpful to provide a date to show when the label was installed as some labels can be inadvertently left in place when the work is finished.

Appendix C
Permit-to-work procedures

A permit-to-work is a formal procedure designed to make employees aware that essential precautions and, where necessary, physical safeguards such as carrying out safe isolation, locking off and installation of mechanical constraints, have been put in place.

Permits-to-work are issued as part of an integrated *safe system of work*. Generally, a permit-to-work is a management procedure in which only persons having specific management authority (authorised persons) can sign a permit allowing activities, such as in the case of electrical installations, safe isolation or switching off for mechanical maintenance, upon which a person's life might depend, to be carried out. A permit-to-work is effectively a statement that measures have been taken such that it is safe to work. A permit-to-work should never be issued on equipment that is still live. As such, a permit-to-work should not be issued until after an installation, circuit, or item of equipment has been safely isolated and, in the case of switching off for mechanical maintenance, all other precautions required to prevent movement of electrically actuated parts have been put in place.

A permit-to-work should be kept as simple as possible, contain clear, concise instructions including simple circuit diagrams or schematic drawings to be followed and should, in the case of a failure to properly implement, be supported by the implementation of appropriate disciplinary measures.

A permit-to-work should include the following:

▶ The person(s) to whom the permit applies (the person(s) who will work on the equipment). In practice one individual may be designated as the responsible person
▶ Identification of the circuit(s) or equipment which has been made dead and its precise location
▶ The points used for isolation (switchgear location and/or reference number)
▶ Where conductors are earthed, where applicable
▶ Location where safety locks have been applied and warning notices have been posted
▶ The nature of the work activity to be carried out
▶ The presence of any other sources of hazard, with a cross-reference to other relevant permits which may have been issued
▶ Details of any other precautions which have been taken during the course of the work.

The person issuing the permit should explain the extent of the work covered to the person receiving it. Both persons should agree the extent of the work covered by the permit before they sign it.

The subject of permits-to-work is covered in greater detail in the HSE publication *Electricity at work. Safe working practices* (HSG85) and is also discussed in the IET publication *Electrical Maintenance.*

A model permit-to-work form and associated guidance is included on the following pages.

Notes on Model Form of Permit-to-Work

(a)   Access to and work in fire protected areas

Automatic control

Unless alternative approved procedures apply because of special circumstances then before access to, or work or other activities are carried out in, any enclosure protected by automatic fire extinguishing equipment:

(i) the automatic control shall be rendered inoperative and the equipment left on hand control. A caution notice shall be attached.

(ii) precautions taken to render the automatic control inoperative and the conditions under which it may be restored shall be noted on any safety document or written instruction issued for access, work or other activity in the protected enclosure.

(iii) the automatic control shall be restored immediately after the persons engaged on the work or other activity have withdrawn from the protected enclosure.

(b)   Procedure for issue and receipt

(i) A permit-to-work shall be explained and issued to the person in direct charge of the work, who after reading its contents to the person issuing it, and confirming that he/she understands it and is conversant with the nature and extent of the work to be done, shall sign its receipt and its duplicate.

(ii) The recipient of a permit-to-work shall be an electrically skilled person who shall retain the permit-to-work in his/her possession at all times whilst work is being carried out.

(iii) Where more than one working party is involved, a permit-to-work shall be issued to the competent person in direct charge of each working party and these shall, where necessary, be cross-referenced one with another.

(c)   Procedure for clearance and cancellation

(i) A permit-to-work shall be cleared and cancelled:
- when work on the apparatus or conductor for which it was issued has been completed;
- when it is necessary to change the person in charge of the work detailed on the permit-to-work;
- at the discretion of the responsible person when it is necessary to interrupt or suspend the work detailed on the permit-to-work.

(ii) The recipient shall sign the clearance and return to the responsible person who shall cancel it. In all cases the recipient shall indicate in the clearance section whether the work is 'complete' or 'incomplete' and that all gear and tools 'have' or 'have not' been removed.

(iii) Where more than one permit-to-work has been issued for work on apparatus or conductors associated with the same circuit main earths, the controlling engineer shall ensure that all such permits-to-work have been cancelled before the circuit main earths are removed.

(d)   Procedure for temporary withdrawal or suspension

Where there is a requirement for a permit-to-work to be temporarily withdrawn or suspended this shall be in accordance with an approved procedure.

There must be a control procedure for the recording, issue and cancellation of permits and this can be done via a site electrical log book. On a large site there may be several authorised persons, but only one such person can be 'on duty' at any time and that authorised person will sign in in the log book and manage the electrical system and the issue of permits etc. during their period of duty. All permits are listed in the log book with their cancellation or suspension.

PERMIT-TO-WORK (front)

1. ISSUE   No. ....

To ....

The following apparatus has been made safe in accordance with the safety rules
for the work detailed on this permit-to-work to proceed:

....

....

Treat all other apparatus as live.

Circuit main earths are applied at:

....

....

Other precautions and information and any local instructions applicable to
the work (Notes 1 and 2):

....

....

The following work is to be carried out:

....

....

....

Name (block capitals) ....

Signature .... Time .... Date ....

5. DIAGRAM (see over for sections 2, 3 and 4 of this permit)

The diagram should show:

    a   the safe zone where work is to be carried out
    b   the points of isolation
    c   the places where earths have been applied, and
    d   the locations where 'danger' notices have been posted.

PERMIT-TO-WORK (back)

2. RECEIPT
(Note 2)

I accept responsibility for carrying out the work on the apparatus detailed on this permit-to-work and no attempt will be made by me, or by the persons under my charge, to work on any other apparatus.

Name (block capitals) ....

Signature .... Time .... Date ....

3. CLEARANCE
(Note 3)

All persons under my charge have been withdrawn and warned that it is no longer safe to work on the apparatus detailed on this permit-to-work, and all additional earths have been removed.

The work is complete* / incomplete*

All equipment and tools have* / have not* been removed

Name (block capitals) ....

Signature .... Time .... Date ....

*Delete words not applicable.

4. CANCELLATION
(Note 3)

This permit-to-work is cancelled.

Name (block capitals) ....

Signature .... Time .... Date ....

5. DIAGRAM (continue below if needed)

C

Appendix

Safety locks

There has been considerable discussion in this Guidance Note about locks and locking devices. There are several locking devices on the market for locking switches and circuit breakers etc., but they are all to be used with a suitable good quality safety padlock.

Safety padlocks are more expensive than standard padlocks as they only come with one key, and there are many key "differs" so it is unlikely that one would ever get a key that undid more than one lock. If there is to be an amount of switching on a particular site it is wise to invest in a good lockout set and keep it under the control of the site authorised persons.

Combination type locks or padlocks should not be used as they are inherently insecure and the combination number can easily be known to several people.

D

Appendix E

Forms of separation for LV switchgear

Regulation 14 of The Electricity at Work regulations states: No person shall be engaged in any work activity on or so near any live conductor (other than one suitably covered with insulating material so as to prevent danger) that danger may arise unless it is unreasonable in all the circumstances for it to be dead; and it is reasonable in all the circumstances for him to be at work on or near it while it is live; suitable precautions (including where necessary the provision of suitable protective equipment) are taken to prevent injury.

Where it is impractical to totally isolate an assembly prior to carrying out installation work within a switchgear assembly, the degree of separation, and the way in which the separation is achieved within the assembly should be considered in a risk assessment to be undertaken by the person appointed to be responsible for the electrical equipment, systems and conductors and any work or activities being designed or carried out on or near the electrical equipment. This risk assessment will consider all relevant factors including:

(a) work to be carried out;
(b) mechanical protection afforded by any insulation and separation;
(c) possibility of initiating a flashover; and
(d) likelihood of an electric shock

The term *"Assembly"* includes floor standing or wall mounting distribution switchboards, panel boards, and motor control centres using electromechanical and/or electronic components. It does however specifically exclude individual devices and self-contained components which control a single circuit i.e. wall mounted starters and fuse switches.

Fundamental Objectives of Separation

The principal reason for separating an assembly is to facilitate access to a part of the assembly whilst other parts may remain energised and in service. Whilst, in general, separation does not improve the electrical performance of the assembly it does provide:

► Protection against contact with live parts belonging to the adjacent functional units;
► Protection against the passage of solid foreign bodies from one unit of an assembly to an adjacent unit. Verification of separation is by application of the appropriate IP test to BS EN 60529 (IPXXB and/or IP2X).

BS EN 61439-2 identifies four main categories of separation; Forms 1, 2, 3 and 4. As a first step users should consider what operations they need to carry out without fully isolating the assembly, and then using a process generally as outlined in the chart below, identify the main category of separation they require for their application.

▼ **Figure E.1** Forms of separation

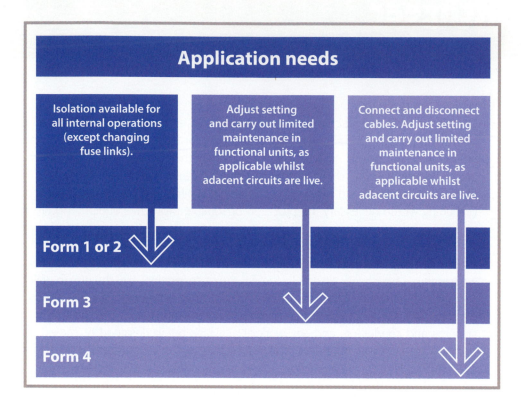

Form 4 provides many variants which offer different opportunities in use. Selecting the most appropriate needs a very detailed understanding of the application and a logical thought process to decide on the functional requirements and to hold discussions with the assembly manufacturer to arrive at the most suitable solution.

Having established the fundamental form for separation needed for an application, specific details must be considered. The IEC and EN versions of 61439-2 divide Forms 2, 3 and 4 into two sub categories. However, within the requirements defined in the standards there are three fundamental ways of providing separation:

▶ Insulation of live parts;
▶ Partitions or barriers (metallic or non-metallic);
▶ Integral housing of a device.

Each has different attributes; clearly much depends on the 'agreement between user and manufacturer'.

In order to aid the agreement between user and manufacturer, BS EN 61439-2 includes a National Annex which further details the sub categories by type of construction such as describing the location of terminals for external conductors and in some instances, the location of cable glands relative to the associated cable terminals.

In making the appropriate choice the user needs to consider:

- The tasks to be carried out with the assembly partially energised – adjusting relay setting, terminating large power cables, replacing components, etc.
- What tools may be used. Is there a risk of tools slipping and damaging insulation.
- Possibility of mechanical impact causing damage to the integrity of the separation.
- Is there a danger of small components falling from one compartment to another causing a hazard.
- Can temporary barriers be effectively used to supplement the protection provided by separation whilst work is being carried out.
- The additional safety that can be provided by the use of Personal Protective Equipment. (This must be seen as a "last resort" and cannot be relied upon for safety in place of a suitable and sufficient design of the assembly to fulfil its intended function).
- The anticipated level of skill of those carrying out any work within the assembly.

In applications where an extremely high continuity of supply is required there may be an advantage in being able to replace or add a functional unit while the busbars and adjacent circuits remain energised and in service. For these applications an assembly with withdrawable or removable functional units should be specified. The level of protection provided to operators while the functional unit is being removed and when it is removed needs to be agreed with the manufacturer.

BS EN 61439-2 is not definitive in what separation will offer or how it will be achieved.

Much is left to agreement between the user and manufacturer. This is intentional as it allows manufacturers to use their initiative, whilst meeting the basic requirements set out in the standard. It also enables the most appropriate assembly to be provided for the particular application, but in so doing, it does make selecting the right assembly all the more difficult.

In addition to the process outlined above the following should be taken into account before reaching a final decision on the form of separation to be specified for a particular application.

This guidance has been taken from the BEAMA Guide to forms of separation for LV switchgear. 2011 edition amended 23 March 2015:
(http://www.beama.org.uk/resourceLibrary/guide-to-forms-of-separation--2011.html)

E

Index

H

I

J

K

L

S

T

U

V

No entries

W

X, Y, Z

No entries

Online training for BS 7671:2018

With the IET's interactive online training from the IET Academy, there's no need to take time away from the job to train for your C&G 2382:18 (or equivalent) qualification.

Book a full or Update course and get:

- **Courses that have been endorsed by C&G, ensuring the best quality content**
- **Learning at your own pace with 12 months access to all content**
- **Online/offline access to suit you, through our app**
- **Three full practice exams to fully prepare you for the real thing**
- **A certificate of completion to present to your local exam centre to sit your C&G 2382:18 exam**

Find out more and book your course now at: www.theiet.org/ academy-regsa